Instrumentation and Control Systems Documentation

Second Edition

Instrumentation and Control Systems Documentation

Second Edition

Frederick A. Meier and Clifford A. Meier

ISA™

Notice

The information presented in this publication is for the general education of the reader. Because neither the author nor the publisher has any control over the use of the information by the reader, both the author and the publisher disclaim any and all liability of any kind arising out of such use. The reader is expected to exercise sound professional judgment in using any of the information presented in a particular application.

Additionally, neither the author nor the publisher has investigated or considered the affect of any patents on the ability of the reader to use any of the information in a particular application. The reader is responsible for reviewing any possible patents that may affect any particular use of the information presented.

Any references to commercial products in the work are cited as examples only. Neither the author nor the publisher endorses any referenced commercial product. Any trademarks or trade names referenced belong to the respective owner of the mark or name. Neither the author nor the publisher makes any representation regarding the availability of any referenced commercial product at any time. The manufacturer's instructions on use of any commercial product must be followed at all times, even if in conflict with the information in this publication.

Printed in the United States of America.
10 9 8 7 6 5 4 3 2

ISBN 978-1-936007-51-6

ISA 67 Alexander Drive
P.O. Box 12277 Research Triangle Park, NC 27709

Library of Congress Cataloging-in-Publication Data is in process.

DEDICATION

This book is dedicated to Cris and Jean,
without whose assistance and encouragement
this book would not have been started,
let alone finished.

ACKNOWLEDGMENTS

We wish to thank all those who assisted in the development of this book, especially Dave Fusaro of *Control*, a Putman Media Co. publication; Ken Brabham of Industrial Design Corporation; co-workers at Harris Group Inc.; and to the Technical and Education Services Departments of ISA, especially Lois Ferson, Alice Heaney and Linda Wolffe; designer, Vanessa F. Harris; and our copyeditor, Jim Strothman.

Cliff would like to especially thank the lads in Dublin, for making it fun, and in particular to Tony Riordan, whose encouragement to have that ceremonial first scoop with "me Da" eventually led to the idea of writing this book.

For the Second Edition, we wish to thank the ISA Staff for all of their help, especially Susan Colwell, Eugenia Bell and Charles Robinson.

Contents

List of Illustrations XI

About the Authors XV

Introduction—Instrumentation and Control Systems Documentation ...1

Chapter 1—The Process Flow Diagram 15

Chapter 2—P&IDs and Symbols 27

Chapter 3—Lists, Indexes, and Databases 69

Chapter 4—Specification Forms 83

Chapter 5—Purchasing 101

Chapter 6—Binary Logic Systems 119

Chapter 7—Loop Diagrams 139

Chapter 8—Installation Details and Location Plans 161

Chapter 9—Drawings, Title Blocks & Revisions 177

Chapter 10—Role of Standards and Regulations 189

Appendix A, Answers to Chapter 2 Exercise 201

Appendix B, Abbreviations 203

Appendix C, Typical ISA-TR20.00.01 Specification Form 205

Appendix D, Drawing Sizes 207

Appendix E, Recommended References 209

Index 213

List of Illustrations

I-1 Typical Continuous Processes . . . 6
I-2 Field Mounted Pressure Gauge . . . 10
I-3 Typical Instrument Identification/Tag Number . . . 10
I-4 Control System Drawing Schedule . . . 13

1-1 Process Flow Diagram . . . 18
1-2 Process Description . . . 19
1-3 PFD Equipment Symbols . . . 21
1-4 Recipe Contents . . . 23

2-1 Instrument & Process Control Defined . . . 31
2-2 Sensing & Comparing Defined . . . 32
2-3 Correcting Defined . . . 33
2-4 Loop Defined . . . 34
2-5 Valve Failure . . . 35
2-6 An Electronic Loop . . . 36
2-7 Identification Letters . . . 38
2-8 Filter Press with D/P Gauge . . . 40
2-9 Filter Press with D/P Transmitter . . . 40
2-10 Succeeding Letters . . . 41
2-11 Instrument Numbering . . . 45
2-12 Intrumentation Devices and Function Symbol . . . 47
2-13 Line Symbols . . . 49
2-14 Pneumatic Transmission . . . 51
2-15 Typical Electronic Transmission . . . 52
2-16 Valve Symbols . . . 53
2-17 Orifice Plate Primary Elements . . . 54
2-18 Measurement Symbols: Primary Elements . . . 54
2-19 Descriptions . . . 55
2-20 Instrument Symbols . . . 56
2-21 P&ID . . . 59
2-22 Flow Loop F-100 . . . 63
2-23 Pressure Loop P-100 . . . 64
2-24 Level Loop L-100 . . . 65
2-25 Local Panel Switches & Lights . . . 66

3-1 Instrument Index 71
3-2 Typical Example of a Company's I&C Data Flow Diagram 76
3-3 Instrument Data Fields 77
3-4 Common Service Description Formats 79

4-1 Specification Forms 84
4-2 Pressure Gauge - Specification Form 88
4-3 Pressure Gauge - Instructions 89
4-4 Level Instrument - Specification Form 92
4-5 Level Instrument - Instructions, Part 1 93
4-6 Level Instrument - Instructions, Part 2 94
4-7 Hazardous (Classified) Locations 96
4-8 Area Classification - Electrical 96
4-9 NEMA Standard 250-2008 99
4-10 Intrinsic Safety 99

5-1 Seven Steps of Acquisition 101

6-1 Instrument Diagram 123
6-2 Functional Diagram 124
6-3 Functional Diagramming Symbols 125
6-4 Electric Schematic Diagram 126
6-5 Logic Diagram Definition. 128
6-6 Binary Logic Symbols - AND & OR. 129
6-7 Binary Logic Symbols - NOT & Basic Memory 130
6-8 Motor Start Circuit and Logic 132
6-9 Logic Diagram L-1 133
6-10 Integrated Control System 135
6-11 Separated SIS 136

7-1 A Pneumatic Loop 141
7-2 An Electronic Loop 141
7-3 Loop Diagram - Terminal Symbols 147
7-4 Loop Diagram - Energy Supply 148
7-5 Loop Diagram - Instrument Action 149
7-6 Loop Diagram - P -100 150
7-7 Loop Diagram - Electronic Minimum 152
7-8 Loop Diagram - Electronic Minimum & Optional 154

8-1 Installation Detail, Type 1 - Thermowell 167
8-2 Installation Detail, Type 2 - Flow Transmitter 168
8-3 Installation Detail, Type 3 - Conduit 169
8-4 Location Plan, Approach A 170
8-5 Location Plan, Approach B 172
8-6 Location Plan, Approach C 173

9-1 ASME Document Sizes for Engineering Drawings. 178
9-2 European Document Sizes 180
9-3 Typical Title Block 181
9-4 Typical Revision Block 185
9-5 Notes and Revision Cloud 187

10-1 Hazardous Area Classification 190
10-2 29 CFR 1910.119 (d) Process Safety Information 191
10-3 29 CFR 1910.119 (f) Operating Procedures. 192
10-4 29 CFR 1910.119 (l) Management of Change 192
10-5 ISA-5 Documentation Series 195

ABOUT FRED MEIER

Fred Meier's career spans more than 50 years as a control systems engineer, chief engineer, and engineering manager in the oil, chemical and engineering industries in the United States, Algeria, Canada, Germany, Japan, and the United Kingdom. He has held Professional Engineer licenses in New York, New Jersey, California, Alberta, Manitoba, and Saskatchewan. He completed U.S. Army training as an electrical engineer and has a Mechanical Engineering Degree from Stevens Institute of Technology and an MBA from Rutgers University.

Fred has been an ISA member for more than 40 years. He has served as President of the New York Section; the Edmonton, Alberta, Section; and the Tar Heel (North Carolina) Capital Area Section. He was awarded the ISA District II Golden Eagle Award in 2000 and ISA's Golden Achievement Award in 2005.

Fred presented two papers at ISA 1982, "Why Not Be an Adaptive Manager?" and, jointly with co-worker Trevor Haines, "Contractor Handling of Engineering for Distributed Control Systems". He authored the cover article for *CHEMICAL ENGINEERING*, Feb. 22, 1982, "Is your control system ready to start up?" Fred and son Cliff (this book's co-author) presented a joint paper at ISA 1999, "A Standard P&ID, elusive as the Scarlet Pimpernel". Fred also published an editorial viewpoint in *ISA TRANSACTIONS*, October 2002, "A P&ID standard: What, Why, How?'. This book won the Raymond D. Molloy and the Thomas G. Fisher Awards in 2004.

After Fred's "first" retirement, he served as the ISA Staff Engineer; after his "second" retirement, as an ISA Instructor and Consultant; and, since his "third" retirement, as co-author of this book. Fred and Jean have been married for 62 years and are the proud parents of four children, four grandchildren, and two great granddaughters, all scattered across the United States and Canada. They currently live in Carolina Meadows, a retirement community near Chapel Hill, North Carolina.

ABOUT CLIFF MEIER

Cliff Meier's engineering experience started back in the 70s with a Bachelor of Science degree in Mechanical Engineering from Northeastern University. His attraction to the widgets and intricacies of Instrumentation and Controls has taken him to three continents and involvement in nuclear and fossil fuel power generation, oil and gas production, chemical, pulp and paper processes, microelectronics factories and even wastewater treatment. Cliff has worked on projects ranging in complexity from a few control diagrams to complex modernization projects and green-field installations entailing thousands of loops and hundreds of schematics. His career started with manual drafting on Mylar sheets and has transitioned to computer-aided design (CAD) where efficient data handling has almost eclipsed the importance of the physical drawings.

While he enjoys the team relationships of design projects, he finds construction, commissioning and teamwork with skilled maintenance technicians, especially the ones with fists of ham, to be almost as rewarding as writing with his Dad.

Cliff is a member of ISA and holds a professional engineering license for Control Systems in Oregon. He and his wife, Cris, have been married more than 32 years and are the parents of the two finest kids on earth - Will and Helen. They live in Beaverton, Oregon, where they can gaze at snow-capped mountains when it isn't raining, which, come to think of it, is not that often.

Instrumentation and Control Systems Documentation

Introduction

This book is written to be easy to read, with many illustrations and little or no mathematics (and absolutely no calculus!). It will be of interest to engineers and technicians, not only in the control systems field, but in many other technical disciplines as well. Control system groups are unique in that they have to coordinate among all the other work groups in a plant, mill, or factory during design, construction, commissioning and operation. This book explains their varied, all-encompassing language. It will also be of value to plant operating, maintenance, and support personnel who are interested in plant design deliverables (the documentation that a design group usually develops).

The engineering design phase of a typical process plant may last from perhaps a few weeks to several years. Once the plant is built it may operate for thirty or more years. Common sense dictates that the documents developed during the engineering phase should have lasting value throughout a plant's operating life.

The purpose of this book is to provide you, the reader, with enough information to be able to understand the documents and the information on them and to use that understanding effectively. It is hoped this knowledge will be useful, not only in existing plants, but also as a basis for a review and reality check on future engineering design packages. Also—dare we say it—the authors hope to encourage effective discussions among the design team, the construction contractor, and the maintenance team that will lead them to agree on the document set that will most effectively meet all their requirements.

Significant material has been sourced from ANSI/ISA-5.1-2009 *Instrumentation Symbols and Identification* (hereafter referred to as ISA-5.1). Material has also been sourced from ISA-5.2-1976 (R1992) *Binary Logic Diagrams for Process Operations* (hereafter referred to as ISA-5.2).

Explanatory material has also been used from ISA publication, *The Automation, Systems and Instrumentation Dictionary, Fourth Edition*, Research Triangle Park, NC: ISA, 2006) (hereafter referred to as *ISA Dictionary*).

First, we need to understand some terms.

Instrument - as defined in ISA-5.1, is a device used for direct or indirect measurement, monitoring, and/or control of a variable, including primary elements, indicators, controllers, final control elements, computing devices and electrical devices such as annunciators, switches and push buttons.

Instrumentation - as defined in ISA-5.2, is a collection of instruments, devices, hardware or functions or their applications for the purpose of measuring, monitoring or controlling an industrial process or machine or any combination of these.

Process Control – as defined in the *ISA Dictionary*, is the regulation or manipulation of the variables that influence the conduct of a process in such a way as to obtain a product of desired quality and quantity in an efficient manner.

System - from the *ISA Dictionary* definition 4, is the complex of hardware and software that is used to affect the control of a process.

This book is about instrumentation and control systems documentation. The book can best be used in an advisory mode. Sometimes the advice is aimed at the control system personnel who are directing the implementation effort and sometimes aimed at the process control personnel who define what is to happen.

The documents we will look at in this book have been developed by industry over many years to efficiently meet the needs of plant design, construction, operation and maintenance. We will look at process control system documents in two ways. First, we will describe them with enough detail to help the reader understand their form and function. For some of these documents, no published industry standard is available to guide the user about their content. The book will therefore describe what the authors believe is a middle path—one that many will accept but, realistically, one that may not be accepted by everyone or in every detail, but what we believe will yield a typical document set.

You may have heard developers of documentation standards say, "My way or the highway" or "There are two ways to do anything, my way and the wrong way." They take this approach from necessity, since a wishy-washy plant standard is not much of a "standard"; it has little value. The authors will not be as dogmatic, since we want you to develop a document set that works for your facility—one that meets your specific requirements. We believe it is appropriate to approach the development of plant documentation standards for your facility democratically—with input from all the parties that have a stake in the product. However, there is a need at some point for autocracy or maybe a "benevolent dictatorship." Once the standards are set—democratically—they must be consistently and properly used. Someone needs to monitor that use and educate users in the acceptable application of the standards. The plant also needs to establish a mechanism for change that controls standards revisions to ensure that all stakeholders review potential changes. The authors urge you, based on painful experience, to control modifications to the plant standards very carefully once a majority of users have defined the plant's documentation requirements. People lose interest in working with a standard that isn't. Rigid control is critical for an effective system. Develop freely; operate rigidly.

The second way we will look at a typical document set is to use a very simple simulated project to follow the sequence by which the documents are developed. There is a logical, time tested sequence to their preparation. Often the development of one type of document must be essentially complete before the development of the next type of document can be started. If the documents are not developed in the right sequence, work-hours will be wasted, since you will have to revisit the documents later to incorporate missing information.

While the sequence is of more importance to those interested in the design process, it is useful for operating personnel to understand how document sets are developed. If for no other reason, this understanding will help ensure that operating personnel modify all the information in all the affected documents as they make changes.

In the authors' experience, there are many different ways to define and document instrumentation and control systems. All the plants that we have seen which used markedly different document sets from the typical set described in this book were eventually built and operated. Of course, some projects ran smoothly, while others seemed to develop a crisis a minute. Some plants were easier to build, and some took longer, but eventually all the plants were completed. Sometimes, the document set's content had a direct influence on how well the project ran, and a smoothly run project is a less expensive project. In our experience, the quality of the document set has a DIRECT impact on the ease of construction, commissioning, start-up and operations.

The use of computers in engineering design now offers many options to better define the work to be performed. Indeed, the new ways available now with linked documents offer attractive efficiency and accuracy that may compel some to revisit the content of the standard design package document set.

You will see that the drawings, specifications and other documents generated in support of process control are unique in many ways. Most control system drawings are schematic in nature, showing how things are connected but not how far apart they are. They are not much concerned with orthographic dimensions; instead they concentrate on the relationships between elements. Unlike with piping plans or structural drawings, the creators of control system drawings are less concerned that elements are X inches from each other and more that the documentation shows the interrelation between field measurement elements and final control elements.

The devices we deal with are becoming infinitely configurable, so our drawings and documents have evolved to capture each device's configuration for reference so the people working with the devices can understand why they were set up the way they are and how they will react to signals.

We use our documents to coordinate with the disciplines that install our control valves. The electrical designers provide power to them, and the mechanical designers provide pressurized air to drive the valves. The control valve specification form is developed by the control systems design group. It describes the flanges on the control valves. The piping group installs matching flanges as part of the piping design and construction.

Our designs must address device failure due to the impact that failure will have on the process, so our documents must state failure action and response parameters for record.

Some instruments require power. This information is defined by the control systems design group but supplied and installed by electrical design and construction.

We generate a lot of documents; it is not uncommon for a set of loop diagrams on a project to be far thicker than all the drawings generated by other disciplines combined (see Chapter 7 for definitions of a loop and a loop diagram). We have input, or draw critical information, from all the other technical disciplines as well, so control systems personnel have an obligation to play well with others: our devices have to meet the connection specifications of the piping group and our components must be appropriate for the assigned electrical area classification, to name a few.

Over the years the authors have noted that people working in process control tend towards an affinity for pattern recognition. Since the sheer volume of devices being controlled and, therefore, recorded on drawings is so great, control systems technicians tend to rely on each one being connected like the others. They understand that the uses vary and the operating parameters may change, but the devices and connections should be the same. Thus, people working in control systems rely on standardization. Without standardization of information and even devices, the work of a control systems technician becomes overly challenging. The standardization criteria may be contrary to competitive bidding goals, but the impact of managing change is ongoing and expensive, and can lead to confusion and perhaps unsafe conditions.

As a technical discipline, our work uniquely bridges between other disciplines. To be truly effective, control systems designers have to be aware of how their work affects other disciplines and how other disciplines affect their work. They have to work closely with those running a facility to ensure that the process control interface presents to an operator the information needed to run the facility effectively. Operator interface designs need to show just enough information to control the process without overwhelming the operator and to provide effective alarms for abnormal states so the operator can respond quickly and effectively.

Types of Processes

Some of the documentation needed for a control system is independent of the type of process, since it exists to define the components and their interconnection regardless of what they are doing. There are basically three types of industrial processes: continuous, batch, and discrete manufacturing. A brief description of each type follows:

Continuous: Material is fed into and removed from the process at the same time. Petroleum refining is a good example.

Batch: A defined quantity or "batch" of material is isolated, and is subjected to a modification; for example, a chemical reaction. The modified material is then frequently subjected to another step, or many. Many repeats of this process, perhaps using different equipment, may be necessary to make the finished product. Beer is a wonderful example of a product made in a batch process.

Discrete manufacturing is defined in the *ISA Dictionary* as the production of individual (discrete) items (e.g., automobiles, electronic devices). Separate components, parts or sub-assemblies are manufactured or assembled to produce a product. Automobile manufacturing is an example.

The "process industry" sector of the worldwide economy consists of plants that operate continuously and those that operate in batch mode. Since there are similarities in design and operation, plants that operate continuously and those that operate in batch mode are generally combined under the "process industries" label. The process industry sector is defined in the *ISA Dictionary* as follows: those processes that are involved in but not limited to the production, generation, manufacture, and/or treatment of oil, gas, wood, metals, food, plastics, petrochemicals, chemicals, steam, electric power, pharmaceuticals and waste material. All the documents discussed in this book are common in the process industries.

The nature of the documentation used to describe modern control systems has evolved over many years to maintain a primary objective: to efficiently and clearly impart salient points about a specific process to the trained viewer. As the processes become more complex, so then does the documentation. An ancient, simple batch process like making brine might be defined quite clearly without so much as a schematic drawing, simply by showing a few pipes, a tank and some manual valves. A modern continuous process that runs twenty-four hours a day, seven days a week, with specific piping and valve requirements, many interrelated controls, and numerous monitoring points, operator control requirements, pumps, motorized equipment and safety systems will, of course, require a more complex documentation system. Figure I-1 shows examples of typical continuous processes.

Figure I-1: Typical Continuous Processes

- Steam production
- Chemical reactions
- Separations
- Waste treatment
- Distillation

The definition of continuous operation from the *ISA Dictionary* adds to our understanding. It reads as follows: "a process that operates on the basis of continuous flow, as opposed to batch, intermittent or sequenced operations."

As the amount of information needed to define the process increases, the documents must become more specialized, allowing for the efficient grouping of details. The piping design group develops and maintains their line lists; the control system design group does the same with their Instrument Indexes. Although both lists are keyed to a general supervisory document in some simple way, the lists themselves are extremely detailed and lengthy, containing information of value to specialists but not necessarily important to others.

General information that defines a process is maintained in a form that is both simple and easily read, but without all of the detailed information needed by a specialist. An example of a general supervisory document is a Piping and Instrumentation Diagram (P&ID). The general document serves as the key to the more detailed documents. Information presentation and storage thus become more efficient. The overall picture and shared information of use to most people are on the general document. Information of use to specialists to flesh out the design is maintained on the detailed documents.

The documents that describe modern industrial processes, like most technical work, assume some level of understanding on the reader's part. The documents use a schematic, symbol-based "language" that may resemble Mayan hieroglyphics to those unfamiliar with the process nomenclature. The symbols, however, provide a wealth of information to those trained to translate them.

Both tradition and standards govern the presentation of these symbols on a document. Indeed, the very existence of some types of documents may seem odd unless the observer understands their intended function. Like any living language, the symbols and their applications are being improved constantly to meet new challenges.

If you have recently entered the profession, this book will train you to read, understand, and apply the symbols and documents used to define a modern process control system. For more experienced professionals, it will offer insights into using the symbols and documents effectively, including explanations for their use. It will present variations that the authors have seen in the use of symbols and documents, and will point out some pitfalls to avoid.

To better understand process design documentation today, in this book we will look at how and when documents are developed, who develops them, why they are developed, and how they are used. The types of documents we will discuss include Process Flow Diagrams, Piping and Instrumentation Diagrams, Instrument Lists or Indexes, Specification Forms, Binary Logic Systems, Installation Details, Location Plans and Loop Diagrams. We also will investigate how these documents can be used to best advantage during plant construction and operation.

The authors are strong proponents of honoring and using standards, including industry standards developed by the International Society of Automation (ISA) and other organizations, as well as plant standards developed especially for and by staff at a specific location. However, we are not zealots. The documentation must fulfill a need and must not present information simply because someone perceives that it is called for by some standard. That said, you should understand that industry standards are almost always more "experienced" than you are. They have been developed, reviewed, and time tested. You should not deviate from any standard unless you have carefully considered all the ramifications of doing so, and have obtained permission to do so from a recognized authority.

The authors know of one large corporation that does not use Loop Diagrams. They have been able to meet their maintenance, configuration, construction, and purchasing requirements with some very creative use of instrument databases. However, they arrived at the stage where they felt confident changing their usual document set after carefully considering and testing some assumptions. They reviewed the proposed document set with all concerned parties, including their design and construction contractors and their own management, before committing to using databases in lieu of Loop Diagrams. That being said, the information they maintain and present in their databases is the very same information contained in a Loop Diagram, without the graphical representation. This is a critical point: the retention and control of the data is still the primary consideration; only the format they chose to present the information was less traditional. Control system documents have to "work" to be effective. Plant design and operations personnel using them must have confidence that the information shown is accurate and up-to-date. A facility might be operating unsafely if there is no culture or system in place for recording changes on the affected documents. If this pipe no longer connects to that piece of equipment, is that associated relief valve still protecting what it should? If not, you might have code compliance issues, not to mention a potential safety hazard. And the best control system in the world will be unable to maintain the process temperature if there is insufficient coolant due to undocumented tie-ins that have depleted the available cooling water.

Changes or upgrades to your facility need to be based on the reality of what is actually installed. If documents are not kept accurate and up-to-date, future work at your facility will be extraordinarily and needlessly expensive. The lack of accurate, current documentation can actually kill a project that otherwise would be economically viable, due to the requirement for and cost of verification. The designer or the construction contractor will have to verify the current condition of the process before implementing changes. An effective change must be made based upon what you really have rather than on what you had or, worse, what you think you have.

The modern industrial facility can be chaotic at times. However, plant and project personnel must be able to communicate easily. An industry-recognized language facilitates that communication. Design projects are difficult enough in today's economic environment without the additional work-hour burden of developing unique instrumentation symbols to define systems when a more recognized and understood system is already available in ISA-5.1. And, believe us; some control system designer, technician, or pseudo expert in your design firm right now may be doing just that. The authors also want to point out that industry standards allow you to make variations in the content of the documentation to suit your specific requirements. ISA-5.1 contains both mandatory and non-mandatory statements. The developers of ISA-5.1 hope this will enhance the strengths and lessen the weaknesses of previous issues of the standard.

The industry standards discussed in this book have been tested over time, and they work. This book will explain how and why they work; it is up to you to apply this knowledge. Of course, the documentation you use and its content must stand the "customer" test. They must be of value to the user; they must be useful! A perfectly executed Loop Diagram with all the features outlined in ISA-5.4-1991 *Instrument Loop Diagrams* is of little value if no one finds the information useful.

The following eight document types—discussed in detail in this book—have been used successfully as a typical set of documents for many years, even back in the Dark Ages of manual drafting on linen or Mylar and the ammonia smell of blueprints.

Process Flow Diagram

The Process Flow Diagram (PFD) defines the major elements of the process schematically. It shows what and how much of each product the plant will make, the quantities and types of raw materials necessary to make the products, what by-products are produced, the critical process conditions—pressures, temperatures, and flows—necessary to make the product, and the major piping and equipment necessary. For a very simple PFD, see Chapter 1, Figure 1-1.The Process Flow Diagram (PFD) is the starting point for designing any process plant. It is the macroscopic, schematic view of the major features of a process; it is the "talking document" for managers, planners and the specialists of a process design team. The control system design group has little involvement in developing the PFD due to its macroscopic nature; however, the PFD may be quite useful to them later when developing operator interface screens on the shared display screens of the control system. Shared display is defined in ISA-5.1 as the operator interface device, a video, light-emitting diode, liquid crystal, or other display unit, used to display process control information from a number of sources at the command of the operator, often used to describe the visual features of a distributed control system, programmable logic controller, or other microprocessor or mainframe computer-based system.

PFDs are used to develop the project scope; they may also be used to document and maintain overall material and energy balances. For any specific project, PFDs are normally issued for the purpose of gathering comment and review. After questions and clarifications are resolved, the general scope is essentially established, and the P&IDs are then started along with the detailed scoping, estimating and design processes.

Piping and Instrumentation Diagram

The Piping and Instrumentation Diagram (P&ID) is the master design document for a process. Using symbols and word descriptions it defines the equipment, piping, instrumentation and indeed, the control system. It is also the key to other documents. For example, instrument tag numbers are shown on a P&ID. The instrument tag number is the key to finding additional information about any specific device on many other documents. The same is true for (pipe)line and equipment numbers. For a P&ID, see Chapter 2, Figure 2-21.

Developing P&IDs is a very interactive process. Specialists designing electrical, control systems, vessels, mechanical equipment and piping, and even civil and structural designers for some processes, all provide input into their development. Each specialist group puts information on the drawing in a standardized way, adding details as they become available. Properly used, the P&ID is the primary coordination document for design, the premier training tool for operations and records the history of the process design of any facility.

We will discuss symbols and tag numbers in greater detail in Chapter 2. Briefly, a symbol defines the type of instrument, and the instrument tag number identifies the device. An instrument tag number consists of a few letters that describe the function of the device, plus a combination of a number and letters that uniquely identify it. There will be more discussion on this later.

See Figure I-2 for an example of an instrument that might be shown on a P&ID. The circle shows a field-mounted instrument located on a pipe. The "PG" further describes the device as a pressure indicator or gauge. In this instance, sequential numbering is used. Since the gauge is the first of its type on the P&ID, the instrument number "1" is added. The next pressure gauge in this numbering system would have the tag number "PG-2". Some tag numbers are much more complex. See Figure I-3 for a very complex tag number: "10-PDAL-01A-1A1." The prefixes and suffixes further define the location of the instrument and are used to maintain the uniqueness of the loop number.

Figure I-2: Field Mounted Pressure Gauge

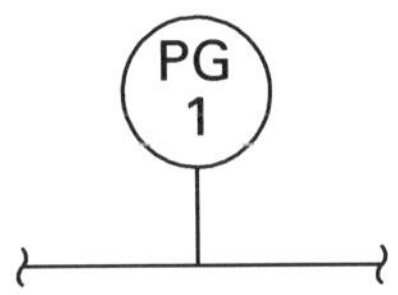

Figure I-3: Typical Instrument Identification/Tag Number - 10 -PDAL-01A-1A1

	10	-	P	D	A	L	-	01	A	-	1	A1	Instrument Identification/Tag Number
												A1	Additional Tag Number Suffixes
											1	1	First Tag Number Suffix
		-								-		-	Recommended Punctuation
									A			A	Loop Number Suffix
								01				01	Loop Identification Number numerals
							-					-	Optional Punctuation
						L						L	Function Modifier letter
					A							A	Function Identification letter
					A	L						AL	Succeeding Letters
				D								D	Variable Modifier letter (if required)
			P									P	Measured/Initiating Variable letter
			P	D	A	L				-		PDAL	Function Identification letters
		-										-	Optional Punctuation
	10											10	Loop Number Prefix

From ANSI/ISA-5.1-2009

Instrument List or Index

The Instrument List or Instrument Index is a list of the data related to a facility's control system components and, possibly, their functions. Instrument Indexes are organized using the alphanumeric tag numbers of the control system devices. They reference the various documents that contain the information needed to define the total installation. Instrument Indexes are discussed in Chapter 3. The terms list and index are essentially interchangeable.

The general term database is also used. It has many definitions in the *ISA Dictionary*. The most simple is: any body of information.

The control systems design group personnel place tag numbers on the P&ID and enter them into the Instrument List or database for tracking. This is done for control purposes because, on a large project, there may be many P&IDs—perhaps one hundred or more—plus thousands of tag-marked devices. Since each device serves a specific function, all devices' status must be tracked until they are installed during construction, their operation has been verified during commissioning, and the plant has been accepted by the owner. Furthermore, each device must be uniquely tracked so its configuration and measurement or control range are known, and many facilities capture the devices' maintenance history as well.

Specification Forms

Specification Forms (or Instrument Data Sheets) define each tag-numbered instrument with sufficient detail that a supplier can quote and eventually furnish the device. For a typical Specification Form, see Chapter 4, Figures 4-4, 4-5 and 4-6. More importantly, the Specification Form retains the critical information needed by control system technicians, such as the manufacturer, model number, range, power requirements and other features needed to define the device for maintenance.

After tag numbers are entered on the Instrument Index or List, the control system design group starts a Specification Form for each tag-marked item. Developing these Specification Forms can be a major part of the control system design group's effort. Specification Forms must be completed to secure bids from suitable suppliers, to purchase the items from the successful bidders, and to generate a permanent record of what was purchased.

Binary Logic Systems

There usually is some on-off or binary or discrete control in a continuous process plant control system. Discrete control is defined in the *ISA Dictionary* as on-off control. P&ID's are excellent documents to define continuous control systems. Other methods are needed to define on/off control. ISA-5.1 and Chapter 6 include descriptions of many of these as does ANSI/ISA-5.06.01-2007 *Functional Requirements Documentation for Control Software Applications.*

As the design progresses, the need to define on-off control will become evident. For instance, on a pulp and paper mill project, it may be necessary to isolate a pump discharge to prevent pulp stock from dewatering in the pipe if the pump is shut down. An on-off valve is added to provide the isolation, but it is necessary to document why that device was added and what it is supposed to do. Since this on-off control may affect many design groups, it is important to define it as early and as accurately as possible.

Loop Diagrams

A Loop Diagram is a schematic representation of a control loop, which in its idealized form is comprised of a sensing element (often called a transmitter), a control component (perhaps part of a shared display, shared control system), and a final control element (usually a control valve or a variable speed drive on a motor). It depicts the process connections, the instrumentation interconnection, connections to the power sources, and the signal transmission methods, whether pneumatic, electronic, digital or a combination thereof. For a typical Loop Diagram see Chapter 7, Figure 7-7.

Finally, when all connection details are known and electrical design has progressed to the point that wiring connection points are known, the control systems design group can develop Loop Diagrams. These diagrams show all the information needed to install and check out a loop. Because these diagrams may repeat information that the piping and electrical design teams included on their drawings, it is critically important that the control systems design group coordinates closely with other disciplines.

Installation Details

Installation Details are used to show how the instruments are interconnected and connected to the process. They are also a primary coordination tool between disciplines. The details provide the means used to mount and support the devices and the specific requirements for properly connecting them to the process. Installation Details are discussed in Chapter 8.

The control systems design group develops Installation Details based on the specific requirements of the devices it has specified, along with any facility owner-driven requirements. The installation requirements needed for good operation and control are established by the instrument suppliers, by various industry groups and by the owners themselves. These requirements are then documented in the Installation Details. These details may be developed for the project, for the specific site, or possibly by the owner's corporate entity.

Location Plans

Location Plans are orthographic views of the facility or process area, drawn to scale, showing the locations of field mounted transmitters and control valves.

They often show other control system hardware including marshaling panels, termination racks, local control panels, junction boxes, instrument racks, instrument air piping or tubing and perhaps power panels and motor control centers. Location Plans are discussed in Chapter 8.

At the same time, the plant layout has also progressed, so the control system design group can begin placing instrument locations on the Location Plans. These drawings are most often used to assist the construction contractor in locating the instruments, but they can also be useful for operations and maintenance because they show where instruments are installed in the completed plant.

Logical Sequence of Document Development

These eight document types are developed sequentially as the project progresses and as the relevant information become available. See Figure 1-4, Control System Drawing Schedule which illustrates typical sequential document development.

Figure I-4: Control System Drawing Schedule

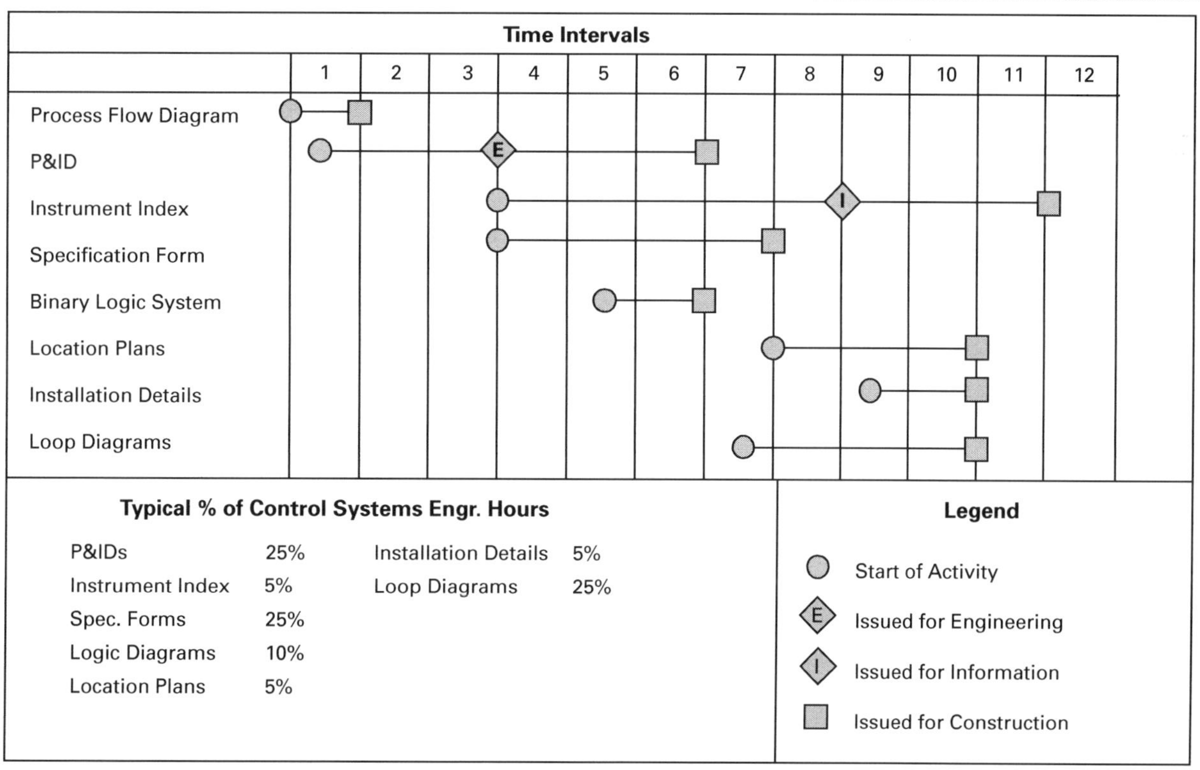

Summary

In this introduction we have briefly described the documents that are included in the control systems set of deliverables and the sequence of their development. In the following chapters we will add more detail to describe the documents, how they can be used effectively, and how industry standards can assist.

Note: Many illustrations in this book were originally developed for various ISA training courses, ISA standards, and other ISA publications. The origins of some illustrations are noted adjacent to the figures. Some of the illustrations were revised for clarity and consistency.

CHAPTER ONE

The Process Flow Diagram

The Process Flow Diagram (PFD) is a specialized document that graphically depicts the key elements of an entire facility or process; but due to that specialization, you may never have seen or used one. It is, nonetheless, a keystone to the organized early development of any complex process and a valuable record of that process for strategic planning. They are invaluable project scope definition tools. These schematic representations are used by corporate managers, strategic planners, process designers and even regulatory agencies because they deal with the larger, macro information: "this much raw material is converted to this much product using this energy in these forms", all on a few neat and tidy drawing sheets and tables. In this chapter we will tend to refer to PFDs in a project or specific process context, but they are used as much to record the function of an entire plant in a concise manner appropriate for purposes that don't require piping and control details.

A PFD is the fundamental representation of a process that schematically depicts the conversion of raw materials into finished products without delving into the details of how that conversion occurs. It defines the flow of material and utilities such as water and electric power; it defines the basic functional relationships between major pieces of equipment. The PFD defines the capability of a system by listing minimum, normal and maximum conditions. They include ranges of flow, pressure, temperature, and possibly some other defining parameter for that process.

PFDs are closely associated with material balances which are used to determine the raw materials and utilities needed to achieve a desired result or product. A material balance is also referred to as a mass balance. PFDs can provide or infer information related to an energy balance as well; as it relates to utilities needed to perform the materials conversion. A material balance in mass units is preferred; engineering calculations at this macro level are typically done in mass units rather than engineering units, that is, pounds per hour rather than gallons per minute.

The material balance starts with a mass calculation performed for a number of conditions; minimum, probably a few "normal" and then maximum conditions. The drawing that ends up as the PFD records the results of those calculations which then serves to define the amount of product a product or facility will produce and the resources and utilities that are needed to achieve that desired result. These calculations don't address the specifics or pipe sizing, or process measurements or controls; instead they focus on how to achieve the end results without addressing the minutia of each step.

Project design teams use PFDs during this developmental stage to document the design options under study. Strategic planners and managers use PFDs to define the capabilities of their facility. Feasibility studies and scope definition work use PFDs prior to commencing detailed designs.

PFDs are not only associated with new construction. Within an operating facility, a plant-wide design group and site management may use PFDs to document the requirements to produce different products or use different recipes and to provide a framework for facility optimization in support of production changes.

There is no generally accepted industry standard available to aid in developing the PFD. ISA-5.1 does define how instrumentation can be depicted on a PFD.

A typical PFD shows the product manufactured or treated by the plant, the raw materials necessary for that product, the by-products produced by the process, the waste materials that must be disposed of, the basic process pressures, temperatures, and flows needed to produce the product, and the major equipment needed. The important piping runs are shown, but piping is not normally sized on a PFD, and auxiliary and utility piping are not shown. A written description of the process may also be included, if only to emphasize certain critical characteristics of the process.

Most PFDs show a minimum of equipment or control detail, while others may include more. These two design approaches are discussed below.

Minimum Detail Approach

For a PFD to be most effective in schematically representing the critical details of a process in the least amount of room, the entire process should be shown in as little space as practical. Only the major process steps are depicted. The intent is to simply show that a change has or will be made to a material or that a product has been modified rather than the minutiae of how that change was made. It can be something of a challenge to determine what should be shown on a PFD, but remember the PFD is a big picture schematic; there will always be a P&ID made to flesh out the details. One should err on the side of removing detail from a PFD.

The PFD content is mostly driven by the customer for the document. It is more in the realm of the process engineer and senior management, and less the playground of the control system design team. To this end, little or no line sizing is provided, there is very little process control shown on a PFD, since this information is not critical to the material balance. Remember that for most projects, process control components are not a significant cost component in the overall budget. Valves and transmitters are usually significantly less costly than an associated pressure vessel. Details will be shown later on the P&IDs and other project documents. P&IDs will be discussed in detail in Chapter 2.

So, what gets shown on a minimum detail PFD? As stated before, the process steps that convert the material from one form to another are generally shown, and the details of that conversion are tabularized in some way, refer to Figure 1-1. A PFD doesn't normally show a flow or pressure control point because these don't change the mass balance. It may, however, show where two streams combine or split to make a third, or possibly where there is a significant utility demand. A successful rule of thumb is to show detail on equipment only if:

- That information has a material balance impact.
- There is a significant utility impact.
- There is a custody transfer point (ownership transfer point).
- Information needed to further define process equipment when the detail design is implemented.
- Special equipment. In practical terms, "special" here probably means "having a significant cost impact on the project."

If the information is needed to reach a critical project decision, it may be important enough to show on the PFD.

Additional Detail Approach

Plant design teams and plant owners may want PFDs to exceed the minimal detail macro approach by including design details that otherwise would only appear on P&IDs. These teams and owners might involve the process control engineers early in the project to gain insight into the cost and implementation challenges of the decisions being contemplated. Those decisions, or the plans that come out of them, might be documented on a PFD if they are seen to be extraordinary, or in other words, costly.

There are thereby varying levels of involvement by control system designers in the development of the PFD. The PFD might include design details such as major measurement points, control methods, some control valves, and some process analyzers. The PFD may be used as a guide, or perhaps even the first steps, in the development of the P&IDs. Details will then be duplicated on the P&IDs and other project documents.

A single PFD may contain enough information for several P&IDs. One rule of thumb is that a single PFD may contain enough information to develop up to 10 P&IDs!

The purpose of the PFD is to define the design of the process. Figure 1-1 is an example of a minimum detail PFD, albeit an overly simple one that only shows a tiny piece of a much larger process. Completion of a PFD is frequently the starting point of the detailed engineering of any process.

Figure 1-1: Process Flow Diagram

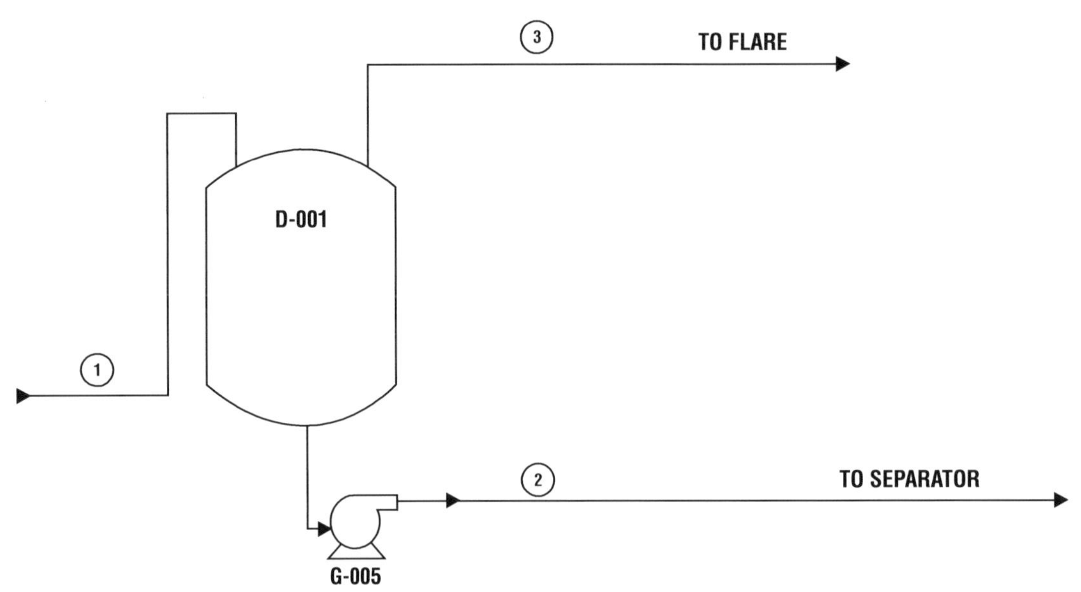

STREAM NUMBER	FLOW	DESCRIPTION	TEMP	PRESSURE	SP GRAVITY
1	10,000#/Hr	WET GAS	90° - 180° F	20 psi	-
2	1,000#/Hr	DEGASSED MATERIAL	70° - 170° F	50 psi	0.9 AT 60°F
3	9,000#/Hr	LIGHT ENDS TO FLARE	80° - 140° F	4 psi	-

ISA COURSE FG15
PROCESS FLOW DIAGRAM
PLANT 001 KNOCKOUT DRUM 0-001
DRG #PFD-1

A PFD is most likely to be developed in several steps. The plant owner may develop a preliminary PFD for their "thinking document" to establish the proposed process or a process change for consideration. The plant owner may also, or instead, elect to use other methods such as a written description to define the process scope. See Figure 1-2, Process Description. In either form, this information is used to establish the initial design criteria for the proposed process or process change.

The PFD, or other conceptual information, is normally reviewed by the owner's process design team and the engineering contractor's process engineers and planning team before detailed design commences. The review is to ensure that two criteria have been met:

1. There is enough information on the PFD to support development of the P&IDs by all the detailed design disciplines. The decision that "enough"

information is presented is probably best left to the design entity or entities that will use the PFD.

2. Material balance information is present to support, with the experience of the project design and purchasing teams, the identification and specification of "long lead" equipment.

Figure 1-2: Process Description

- **Process Description Plant 001 Knockout Drum D-001**
  - The inlet gas, which consists of mixed petroleum liquids and vapors, originates in various sections of the plant and is piped to the knockout drum, D-001, where liquids and vapors are separated by expansion and a slow-down of velocity.
  - The mixed petroleum liquids are pumped to the separator and vapors are routed to the flare.
  - The incoming material is normally 10% condensate, but under some conditions, condensables may be reduced substantially.
  - The wet gas will vary in temperature from a low of 90°F to a high of 180°F.

Long Lead

"Long lead" is a term used to describe equipment or components that take a long time to procure, design, fabricate and ship. In other words, it is equipment that has to be purchased early in the project. The criteria for "long lead" should be defined by the construction management or purchasing groups, since it changes between industries and projects. A long lead item might be one that will consume one-third of the construction schedule between issue of the specification for bid and delivery to the site. If the equipment needs an atypical length of time for installation, it may also be appropriate to call it long lead regardless of the length of the procurement cycle.

The Early Design Effort

The owner may put a lot of effort and invest a great deal of time, money and expertise in a project before any PFD is developed. The following is a simplified look at the early steps the owner or any design team might take.

A project may start with a gleam in someone's eye or a voice in the middle of the night: "We could sell a lot more product if we had a new, more efficient plant. We could sell a new product like soap, paint, sodium bicarbonate, tissue, toluene di-isocyanate, or computer chips provided we could produce it in a cost-effective way by using a new process, new materials, different techniques or even an entirely new plant. We could make our product better, or cheaper. We need to reduce pollution, or have fewer by-products. We could make our product more profitable with higher quality." The gleam in the eye or the voice in the night is then turned over to a group of experienced personnel for further development.

If the proposed project is a new plant, the group will include company managers and specialists such as consultants, engineers, real estate advisors, purchasing managers, marketing teams, sales experts, and other support personnel. As the concepts mature, the group will evolve into teams developing, at the least, a general size and location for the plant, a marketing plan for the product, and a financial plan to establish and control costs. A preliminary process is then defined with a PFD, and the source and costs of the raw materials are determined.

If all this information is favorable, company executives may decide to proceed with the concept. This conceptual plan, grounded in part by the process defined by the PFDs might specify that the plant be located where raw materials, electricity, water and a suitable labor force are available. The economic portion of the plan will have the costs defined based in a general way on the content of the PFDs as well as the experience of the management and designers with cost escalation calculated for the project's duration. The economic plan will include the production yield forecasts as well as the planned cost of the raw materials, combined and massaged to provide a unit cost and margin for the units of product sold. Ultimately, the project will be able to estimate the return on investment (ROI) for the project, which hopefully will be above the company threshold for new projects. If the ROI is below the company threshold, the project is simply not going to be approved.

Planning continues after the decision is made to proceed with the project. Next, the executive team secures the necessary land, and a set of scope definition documents is completed. While individual equipment sizing is not part of a PFD, experience and technical acumen can estimate the amount of land needed by the process shown. This information serves as the starting point for the detailed engineering. When this initial work is done by the owner prior to involvement by a detailed design team, the initial or preliminary PFD, and other process descriptions developed by the owner is included in these scope documents. Many owners use independent engineering contractors for the detailed engineering. Other firms have in-house capabilities and staff and prefer to do the detailed engineering design themselves. The "who" is not important. The process of figuring out what to do, how to do it, what it will cost and the confidence that it will achieve the production and cost objectives are "what's important."

The Design Team

Whether a contractor develops the design, or it is done in-house, the work is done by an engineering design team, consisting of many specialty groups. A typical team will be led by a project engineer or engineering manager and it might consist of the following design groups:

Civil	Process
Electrical	Project
Process Control	Structural
Mechanical Equipment	Vessels
Plant Design/Piping	

The design team is a part of the total organization necessary to manage the design and construction of a facility. One common term for the scope of the total organization is EPC: Engineering - Procurement - Construction. Some owners hire contractors for some or all of the three parts, while others handle all three themselves. The owner's project manager has overall control of the project. The project manager may also have additional staff to handle other functions, such as cost engineering, estimation and legal. Contractors may also use a project manager to control their portion of the project, if they have responsibilities other than engineering.

When an independent engineering contractor is to be used, the owner will use scope documents such as their PFD to aid in securing the contractor's services through competitive bidding or some other selection processes.

PFD Details

The PFD should use symbols and letter designations to identify the equipment on the PFD. It is not necessary to add much detail, as previously discussed. A simple line sketch or even a box will serve to show a piece of equipment. For instance, a heat exchanger can be shown as a simple line representation of a main process flow and a heat transfer medium flow, without implying a particular type of exchanger. For a PFD, the only information needed is that a piece of equipment transfers heat at that point, rather than showing the specific transfer mechanism.

Some projects might identify equipment by using the Symbol Mnemonics shown in Figure 1-3. This particular owner used the letters VSSL for vessels,

Figure 1-3: PFD Equipment Symbols

Subgroup: Process
Symbol Name: Vessel
Symbol Mnemonic: VSSL
Description: A vessel or separator. Internal details may be shown to indicate type of vessel. Can also be used as a pressurized vessel in either a vertical or horizontal arrangement.

Subgroup: Process
Symbol Name: Distillation Tower
Symbol Mnemonic: DTWR
Description: A packed or trayed distillation tower used for separation. Packing or trays may be shown to indicate type of distillation tower.

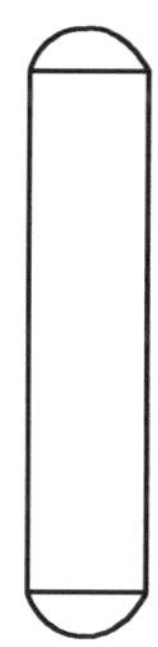

Subgroup: Storage
Symbol Name: Atmospheric Tank
Symbol Mnemonic: ATNK
Description: A tank for material stored under atmospheric pressure.

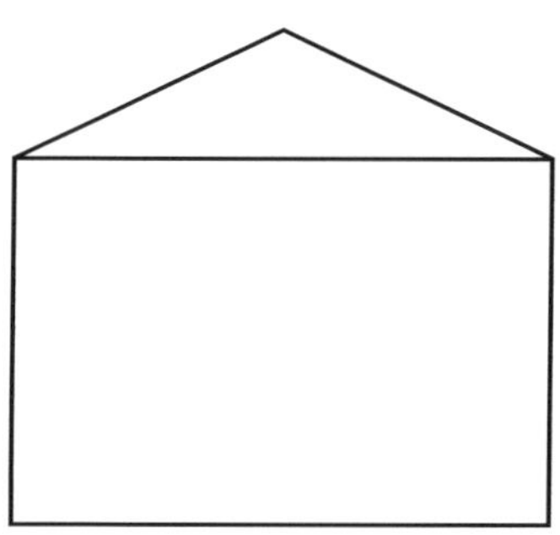

Subgroup: N/A
Symbol Name: Exchanger
Symbol Mnemonic: XCHG
Description: Heat transfer equipment. An alternative symbol is depicted.

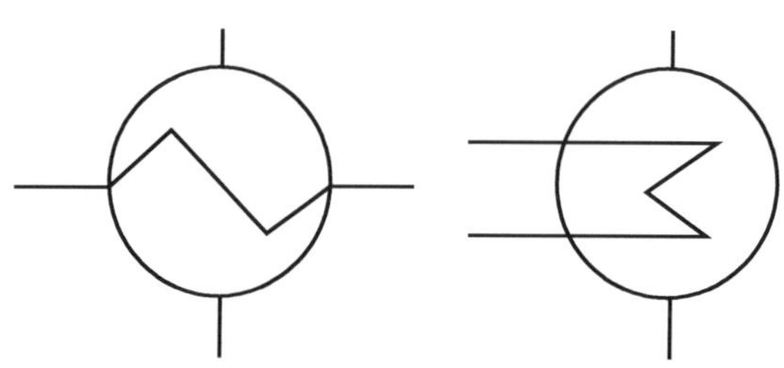

From ISA-5.5

DTWR for distillation towers, ATNK for atmospheric tank, and XCHG for exchanger. Other projects might use a single letter for identification, such as C for columns, T for tanks, D for drums, V for vessels, E for heat exchangers and coolers. We know of more than one project that counter intuitively used G for pumps. Their logic was that pumps are rated in gallons per minute units and they already used P for "pressure vessels". The point here is that people will use terms and symbols with which they are comfortable; our only interest is that they use the same terminology everywhere and that they provide a legend explaining the method. While there are many variations of the letters and symbols used, it is very important to be consistent throughout a project, and almost as important to use symbols familiar to those who will use them.

Process flow data and conditions are tabularized on the PFD. These conditions define the "design" conditions, but—since it is important to the equipment sizing—in addition to normal or operating conditions, the maximum and minimum conditions will be provided as well. Remember, since the PFD is tied closely to the material balance, mass flow units are normally used. Basic pressure and temperature conditions may be provided as well. The mass flow units will be converted to engineering units during the detail design.

There are two common ways to show the process conditions. One is to provide a set of numbers along the line connecting equipment using a standard format, possibly three conditions with slashes for delimiters such as: flow/pressure/temperature. Units are not normally provided to conserve space. The units are standardized and are explained on the legend sheet. The flow conditions are those upon which the project will be based; the conditions on the PFD will eventually become part of the equipment specifications, frequently after the mass units are converted to more common engineering units.

Another useful way to document process conditions is to use a keyed table. A numbered symbol—frequently a diamond with an internal number—is added above a line or piece of equipment on the drawing. A table is then provided along the top or bottom of the PFD listing the process conditions for that numbered symbol. This approach has the advantage of simplifying the addition of process conditions, and makes it a bit easier to maintain the data in the table.

As discussed earlier in this chapter, some engineering contractors or plant owners include more information on PFDs than the minimum. This should be agreed upon between the owner and the contractor before the documents are prepared. Arguably, when there is pressure to add more detail to the PFDs, it may well be time to redirect the design effort to P&IDs. Some projects may show basic measurement and controls information on a PFD because they deem it important to understanding the process shown on the PFD. We don't encourage this approach, but it happens; only very simple, easily recognizable and simple symbols are used.

Batch Processing Plants Vary

Batch processing plants may contain equipment used in different ways, in different sequences - often for many different batches or products at one time, or at different times.

The PFD defines a continuous process very efficiently. Batch processing, however, may require additional definition. A batch process subjects a fixed quantity of material (a batch) to one or more process steps in one or more pieces of equipment. The process takes place in a set of equipment defined in ANSI/ISA-88.00.01-2010, *Batch Control Part 1: Models and Terminology* as a process cell.

The process cell may be used to make a single product or many products. There are two further choices if the cell is making many products. The cell may use different raw materials with different process parameters and either use the same equipment or, alternatively, use different equipment. Many process cells have the capability to process more than one batch of the same, or different, products concurrently. A single PFD can define one process. In batch processing the PFD is often supplemented by a recipe, due to the complexity. Recipes contain five categories of information, as indicated in Figure 1-4, and are specific for the end product.

Figure 1-4: Recipe Contents

Header	Administrative information and a process summary
Equipment Requirements	Information about the specific equipment necessary to make a batch or a specific part of the batch
Procedure	Defines the strategy for carrying out a process
Formula	Describes recipe process inputs, process parameters, and process outputs
Other information	Product safety, regulatory, and other information that doesn't fit in the other categories

Figure 1-4 is from the book written by Jim Parshall and Larry Lamb. *Applying S88, Batch Control from a User's Perspective* (Research Triangle Park, NC: ISA, 2000) page 48. The book contains the definition of a control recipe: "A control recipe is used to create a single specific batch.... control recipes unique to individual batches allow product tracing or genealogy to occur."

ANSI/ISA-5.06.01-2007 *Functional Requirements Documentation for Control Software Applications* also provides insight on this subject.

Engineering designers and owners may use the PFD as a first step in designing the measurement and control system. Important process monitoring and control requirements are then captured as they become known during design development. In this situation, the process design team may indicate where various process variables are to be measured. For example, a circle with a single letter P inside might be used to signify that the pressure at this point is important to the process and should be measured. Likewise, the use of F for flow, L for level, or T for temperature in a circle would indicate where these variables should be measured. Everyone involved should understand that the process measurement and control points shown on a PFD are by no means the only points, they are just notable points identified during project scope development. Many more process measurement points and control points will be defined later and shown on a P&ID.

Some designers or owners might elect to show expensive process control system components on the PFD. For example, an in-line process chromatograph may appear on the PFD, due as much to its cost as its importance to the overall process. Measurement and control points required for regulatory reasons or to standardize metrics across different facilities within the same organization may be added to ensure that the detailed design team includes them in the final project. Project teams may elect to define process variable sensing points, controllers and control valves even more stylistically than that on a P&ID since PFDs are intended to provide a canvas for the broad-brush "artistry" or possibly the shorthand of the process engineers. Remember, though, that the detail required for a complete process control design should be left to the P&ID.

The PFD for our simulated project is shown as Figure 1-1 and a word description of the process is shown as Figure 1-2. No symbols are shown on our sample PFD; showing no instrumentation symbols is a more common approach, by far. Simple symbols without numbers are sometimes shown on PFDs to describe process control devices. A very simple continuous process has been chosen for our discussion; the rest of design documents for the process (and, by extension, the plant) will be developed in the following chapters.

The PFD in Figure 1-1 shows there is a flow in the process line, stream number (1), of 10,000 pounds/hour of wet gas with a temperature between 90°F and 180°F and a pressure of 20 psi. The variation in temperature is caused by process changes upstream of this PFD. Note that only a stream number, (1), (2) or (3) identifies the pipelines. Nothing has been included on line size, material of construction, or pressure class (ANSI 150, ANSI 300, etc.) for any of the piping. Notice that there are no symbols or data shown for the pump driver (an electric motor, most often); only the equipment number, G-005, identifies the pump. This minimalist approach is appropriate at this point since many detailed design decisions will follow after the PFD is agreed upon. Restating points raised earlier, one might see driver details if the driver met other uniqueness criteria such as long lead or high cost, or if the driver was part of a different process cycle, such as a pump driven by a steam turbine.

The wet gas goes into D-001, the Knockout (KO) Drum, where the liquid condenses out of the wet gas stream as the gas expands and cools. The liquid is pumped to a separator (on another PFD) where the water and process liquid are separated. Stream number (2) shows that pump G-005 has a discharge pressure of 50 psi. The pumped liquid has a specific gravity of 0.9 at 60°F. The pump has a capacity of 1,000 pounds/hour and the temperature of the degassed material varies between 70°F and 170°F.

The light ends or gases, flowing at 9,000 pounds/hour and shown as stream number (3), are piped to a flare, which is shown on another PFD. The pressure needed to move this quantity of gas to the flare is 4 psi.

From this simple simulated PFD we have enough information to start development of the P&ID. To the project design team, the PFD becomes less important as the P&ID develops and the process temperatures, pressures, and flow rates are used to develop design criteria. However, if the PFD is kept current as the project develops, it may be used to familiarize contractor and owner personnel with the process. It is usually far easier to understand the basics of a process from a PFD than from the P&IDs.

CHAPTER TWO

P&IDs and Symbols

Overview

The acronym "P&ID" is widely understood within the process industries as the name for the principal document used to define a process—the equipment, the piping and all of the control systems components. *The Automation, Systems and Instrumentation Dictionary*, Fourth Edition defines a Piping and Instrumentation Drawing (P&ID) by describing its function: P&IDs "show the interconnection of process equipment and the instrumentation used to control the process." The fact that the P&ID is the principal, defining document is proven by its widespread use across most processes and industries. Once you become familiar with the "language" of the symbols and the presentation, you will come to appreciate its efficiency and simplicity in documenting salient information in an easily understandable way.

Notwithstanding the ubiquitous nature of the P&ID, you may experience confusion when trying to decipher unique symbols or other depictions on your drawings. You are not alone. This book is intended to help resolve the confusion. The fact that confusion exists is understandable because, oddly, there is no universal standard that specifies the information that should be included on a P&ID or how it should be shown. Even more strangely, the meaning of the letters P&ID are not even universally agreed upon. You may know what the "P" stands for, or what "D" means or even what a P&ID contains, but the person in the facility down the road probably doesn't agree in every way. For instance, the "P" in P&ID may stand for Piping or Process. The "I" may refer to Instrument or Instrumentation. The "D" may mean Drawing or Diagram. P&IDs may even be called Flow Diagrams, which are not to be confused with the Process Flow Diagrams discussed in the previous chapter. P&IDs are also sometimes called Flow Sheets, a term often preceded by the department that initiated or developed them, like Engineering, or Controls, or some other descriptor. In this book, for simplicity, we will refer to the document by the acronym, P&ID; you may define it as you wish.

As mentioned above, there is no universal, national, international or international multi-discipline standard that covers the development and content of P&IDs although an ISA Standards Committee is currently working on such a standard based on Process Industries Practice (PIP) PIC 001, which will be known as ISA-5.7. (More on PIP in Chapter 10.) However, much of the information and use of a P&ID is covered by ISA-5.1 which is an excellent document that defines primarily instrument symbolism. Equipment based symbolism used in a P&ID then follows the method used by ISA-5.1 in deriving standard drawings to represent the family of equipment types with as simple a sketch as possible.

One P&ID commonality is that sets of symbols are used and connections are shown between the symbols to represent the process elements and piping. The symbols represent mechanical equipment, piping, piping components, valves, equipment drivers and instrumentation. These symbols are assembled on the drawing in a manner that schematically and clearly defines the process in the correct process order.

P&ID instrumentation symbols are generally based on ISA-5.1. Additional information is also shown on the drawings to meet the specific requirements of the many different stakeholders that use the drawings, and therein lies the confusion.

This book uses ISA-5.1 as the definitive reference. The authors are aware that this document is newly revised and that future changes can be anticipated, but we are sure the intent and focus of the standard will be maintained.

Another professional organization, Process Industry Practices (PIP), has developed and published many recommended practices. Among these is one on P&IDs. There is additional information about PIP in Chapter 10.

The existing P&IDs in your facility have probably been produced and revised over many years by many different developers. Hopefully, many different individuals have documented revisions to the content—and even the symbolism—of your P&IDs to reflect process improvements and additions, as well as changing control technology. However, unless your company has been incredibly fortunate in maintaining site standards, some of your P&IDs will use symbolism and formats that differ from the original and even from each other. As you probably well know, inconsistent symbolism and formatting of your P&IDs can be annoying or confusing, and more importantly, can make the information they contain subject to misunderstandings.

New P&IDs can be a different story. Although the P&ID is the overall document used to define the process, as discussed in Chapter 1 the first document developed in the evolution of a new process design is often the PFD, the Process Flow Diagram. Once a PFD is released for detail design, the project scope has been established and P&ID development can commence.

It is important at this stage, before P&ID development gets underway, for the facility owner to define their standards and requirements for P&IDs as well as those for other documents. The documents' operational and maintenance needs are, in the authors' experience, not likely to be met by the design team without clear and concise instruction. This instruction is probably developed best in a workshop format with examples provided by the owner so the design team understands what is being requested. The workshop approach is best because each stakeholder needs to discuss and work out how the information on the drawings will be used and how a change in "what we always do" will impact other activities. Establishing these requirements before releasing the design team to begin the work is orders of magnitude more efficient than waiting until the 30% review cycle before discovering, for example, that the owner requires a specific string of letters and numbers to identify equipment within their computerized Asset Management System. At this point in the project, many other documents will be impacted by this "simple" change, and almost every drawing that has been started will have to be revised.

For example ISA-5.1 identifies nine different loop numbering schemes. Some of these are parallel – a duplicated numerical sequence for each loop variable. Other schemes are serial, a single numbering sequence for all loop variables.

One critical element that must be agreed upon before a new P&ID is started is the P&ID legend sheet. This drawing defines the symbols, line types, line identification system, equipment callouts, and acronyms that will be used on the P&IDs. The legend sheet is useful as the starting point for discussion in the P&ID kickoff workshop. The legend sheet is discussed in more detail later in this chapter.

P&IDs develop in stages. The key members of the design team—perhaps plant design, piping, process, and project specialists and the owner, all lay out a conceptual pass showing vessels, equipment and major piping. The instrumentation is typically added next, since it often requires significant space on the P&ID. Or, in the words of one project manager, "You guys sure do have lots of bubbles." Then, the contributions of the specialists in electrical, mechanical equipment, vessels and other disciplines are added. These specialists fill in the information blocks containing equipment numbers, titles and definitive text reserved for critical information regarding the equipment: size, rating, throughput, and utility demand (horsepower, kilowatts, gallons per hour, etc.) The developmental process is an iterative one. Information is added in steps until the document is complete with all necessary details.

P&IDs are controlled documents that are formally issued at various stages of project design. They are considered "milestone" documents in a formal design contract since their completion status directly reflects the project design's "percent complete" for payment purposes.

The term "controlled document" means that changes to the drawings are identified and clearly documented in some manner and that there is verification checking or some other quality assurance procedure in effect. This change documentation is needed because many different design entities have based their work on the content of the prior issue of the drawing; changes need to be called out so that the subsequent design steps can be modified to incorporate those changes.

The owner's organization needs to be vigilant in controlling the content of P&IDs. Since these drawings are the definitive resource used by operations and maintenance staff to understand the process, they are likely to be the document that organizes the plant's equipment identification system and they are the key to work done by most design entities. Consequently, it is necessary to first ensure that all required information is shown and that it is presented in the best way possible. The owner gets to define "required" and once that is established, everyone involved must be sure the expectation is met every time. "Best" in this case is achieved when the equipment shapes, symbols, text, line type, line weight, and content all appear the same way every time so people can read and understand the content at a glance. This is not the place for creative experimentation.

From the P&ID comes the Instrument List or Index, which documents the specification, acquisition and installation of all the instruments. From the P&ID comes the motor list and horsepower. From the P&ID come the piping line list, sizes, service and purpose. The P&IDs even documents critical information regarding tanks, vessels and other equipment. All of this information is used to lay out equipment on Location Plan drawings and to start the specification and purchasing efforts.

In some states, P&IDs carry Professional Engineers' stamps. This means that an engineer licensed by the state where design will be implemented is in charge of the design and will review or approve the drawings as issued. The engineer whose stamp appears on the drawing is responsible for the content and accuracy. This can be a challenging requirement to fulfill when designs are developed remotely from the physical construction site, as is often the case. States issue engineers' licenses independently, so the specific person in charge of a design may have to go through the licensing process in your state, which takes time. Licenses are often not readily transferable from one state to another.

P&IDs are distributed to members of the project team and to interested owner personnel after quality control checking and under rigorous revision control. This formal issue process occurs several times in the course of a project so that all the design entities can work and progress incrementally, rather than waiting until the process is completely defined and having to scramble at the end of the design stage. As mentioned above, these drawings are so important that key milestones are often built into the project schedule based on the different issues of P&IDs. Typical formal P&ID drawing issues may include:

A – Issue for scope definition

B – Issue for client approval

C – Issue for bid; bidding of major or "long lead" equipment

D – Issue for detailed design

0, 1, 2, 3 etc. – Issue for construction

Before we start looking more closely at a P&ID we will define a few terms.

Figure 2-1 contains a few simple definitions. An instrument is a device for measuring, indicating, or controlling a process. This includes both simple and complex devices. Pressure gauges or dial thermometers are typical simple ones. Complex devices may include process analyzers—perhaps a gas chromatograph, which defines the types and quantities of gases in a process stream.

Figure 2-1: Instrument & Process Control Defined

- **Instrument**
  - A device for measuring, indicating, or controlling
- **Process control**
  - All first-level control – process or discrete – consists of three parts:
    - Sensing
    - Comparing
    - Correcting

First-level contol is the control system needed for normal plant or process operation. ISA-5.1 uses the term Basic Process Control System (BPCS).

The term "Process Control" can be understood from any dictionary definition of the two words. In its simplest form, a process is a series of steps and control is to regulate. So process control is the regulation of a series of steps.

ISA-5.1 uses the term Basic Process Control System (BPCS) as the control system needed for normal plant or process operation and High Level Control System (HLCS) as a system above that of BPCS. Safety instrumented systems (SIS) are defined in the *ISA Dictionary* as those systems whose purpose is to take the process to a safe state when predetermined conditions are met. (See Chapter Six for more information on safety instrumented systems)

All types of process control include three functions: sensing, comparing and correcting. Instrumentation, or "measurement and control devices," is used to accomplish each of these functions or even all of these functions simultaneously, along with indicating—presenting information to an operator.

Sensing

First, we have to know where we are by sensing the relevant characteristics of our environment—otherwise known as the process. One definition of process sensing is "to ascertain or measure a process variable and convert that value into some understandable form." (see Figure 2-2).

The flow of liquid in a pipe or air in a duct, the level of liquid in a tank, the pressure of gas in a vessel, and the temperature of the fluid inside a distillation tower are all process variables. Normally, for process control, these variables are measured continuously (certain specialized variables may be measured on a sampling or scheduled basis). In all but the simplest systems, in which the variable is displayed at its point of measurement, a transmitter measures the process in some way and transmits the information to a location where comparing takes place. Some instruments combine two or even three functions in one housing. For example a pressure regulator might sense the pipeline pressure, compare it with the set point or set pressure and control the pipeline pressure all in one housing.

The simplest instruments permit direct reading of a process variable in the field (that is, the process area of the plant). These devices include pressure gauges, thermometers, level gauges and rotameters. A rotameter or variable area flow meter is defined in the *ISA Dictionary* as a variable-area constant-head indicating or transmitting type of rate-of-flow volume meter in which fluid flows upward through a tapered tube, which lifts a shaped plummet to a position such that upward fluid force just balances the weight of the plummet. Next in complexity we see the measurement transmitted remotely, perhaps to a control panel, to be viewed by a person who uses their training and experience to compare the signal against expectations and to respond manually as needed. At the next level of complexity, and arguably the most useful approach in a modern process control system, is to offload that responsibility for action from the operator to a pneumatic or electronic controller or a shared display-shared control system. Shared Display – an operator interface device that is used to display signals or data on a time-shared basis. Shared Control – a controller which permits a number of process variables to be controlled by a single device. All of these instruments and signals are shown on a P&ID.

Comparing

Figure 2-2 contains a formal definition of the comparing function. The value of the process variable is compared with the desired value, called the set point, after which action is taken to bring the two together. Unless the system has been put in manual control, the control is automatic and is (usually) continuous. Comparing takes place in a pneumatic or electronic controller or via a shared display-shared control system, such as a distributed control system (DCS), a programmable logic controller (PLC), a computer chip embedded in a field instrument, or even a desktop computer.

Figure 2-2: Sensing & Comparing Defined

- **Sensing**

  To ascertain or measure a process variable and convert that value into some understandable form

- **Comparing**

  To compare the value of the process variable (PV) with the desired value set point (SP) and to develop a signal to bring the two together. The signal depends on:

  - How far apart the PV & SP are
  - How long they have been apart
  - How fast they are moving toward or away from each other

Comparing used to take place in a control room (or, for purists, in the rack room where the process control computer was located). Now, digital devices and bus technology allow the control comparison algorithms to reside almost anywhere: within the field measurement device, the final control element or back in a more centralized location. Regardless of thc location, the important issue is that the measured process variable is compared against set point. Set point is defined as the input variable that sets the desired variable of the controlled variable.

When comparing, both electronic controllers and digital devices we may look at three characteristics of the process variable:

P – Proportional or gain—how far away the process variable is from the set point

I – Integral or reset—how long the process variable has been away from the set point

D – Derivativc or rate—how fast the process variable is changing

It is just coincidental that the three components of a process control algorithm use the same three letters (PID) as the primary design drawing that details the process under control (P&ID).

Correcting

The control device then develops a signal to bring the process variable and the set point together. This signal is transmitted to a field device that changes the value of the process variable. The field device is referred to as a final control element. This device is most often a control valve or a variable speed pump drive, although there are many others. See Figure 2-3.

Figure 2-3: Correcting Defined

- **Correcting**
  - –To bring the process variable closer to the set point. This is accomplished by the final control element – most often this is a *control valve*
- **Control valves, usually, but not always:**
  - – Are pneumatically actuated, often by a 3-15 psi signal
  - – Can be moved directly by a pneumatic controller
  - – Are actuated by a transducer if the controller signal is electronic or digital

The Control Loop

The basic set of instrumentation for automatic control is made up of three devices—the transmitter that senses and transmits a process variable, the controller that compares it against an expectation, and the final control element, a device that corrects or manipulates the process. These components are interconnected to form a control loop. The interconnection may be pneumatic, electronic, digital, or more commonly a combination of all three. The signals generated or used by instruments have been standardized around the world to simplify maintenance and manufacture; a pneumatic signal is typically a 3–15 psig instrument air signal (psig means pounds per square inch gauge in the United States and in U.S.-influenced industries). If the interconnection is electronic, a 4–20 mA (milliampere) direct current signal is usually used. Other signal levels are sometimes used, notably 6-30 psig pneumatic when additional power is useful for pneumatic valve actuator diaphragms, and 1–5 v DC. and 10-50 mA. The signal level is a function of the control system selected and the components used.

As yet, there is no agreement in the process industry on a single digital transmission standard, and entire books have been written on the relative merits of the various protocols, but progress is being made. It is at least arguable that eventually technological advances and market forces will converge into fewer, more universal standards. For now, the digital transmission system or communications protocol you use is likely driven by edict from your parent corporation or it may be one that is common for your industry. If you are lucky, it will be one that is supported locally, or maybe it is simply the last one used by your design consultant.

Figure 2-4 shows a pneumatic loop controlling the pressure in a pipeline. The loop number is 100, so all the devices in the loop will have the number 100: the process transmitter, controller and final control element. The double cross-hatched lines indicate that information is transmitted pneumatically from transmitter PT-100 to indicating controller PIC-100, and from PIC-100 to control valve PV-100 with a signal varying from a low of 3 psig to a high of 15 psig. The control valve moves according to the value of the 3–15 psig signal. It has an FO identifier, meaning that if the primary power source to the valve is lost, in this case pneumatic pressure, the valve will Fail Open.

Figure 2-4: Loop Defined

A combination of interconnected instruments that measures and/or controls a process variable

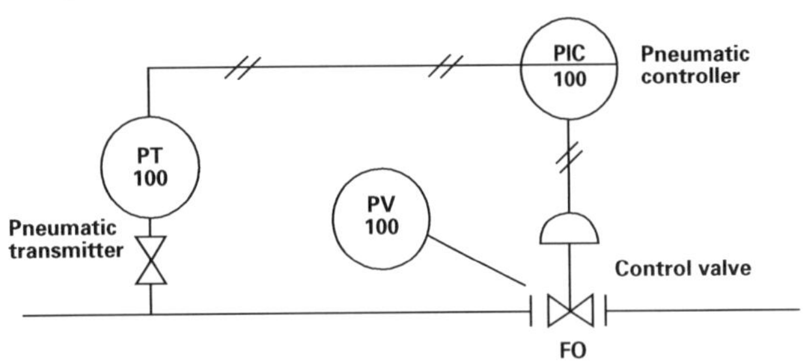

A pneumatic loop - controlling pressure

Control Valve Failure

Control valves may fail in various positions—fail open, fail closed, fail locked in last commanded position, fail in last position drift open or fail in last position drift closed. The position of a failed valve can have a significant impact on associated equipment and on the process and, therefore, it is of great interest to operations personnel. Valve fail action is often discussed and agreed upon during the P&ID review meetings, so it is natural and efficient to document the agreed-upon action on the P&ID. In connection with valve fail action, the

What's Missing?

Is the drawing of the simple pressure loop complete? There probably is no right answer to that—other than, "What do you think?" We are not really ducking the question. But remember, the people responsible for the P&ID will have to live with the drawing for many years. The stakeholders in the project need to decide how much detail is provided on a P&ID. The intended uses of the P&ID as a design document and construction document and to define the system for operations all will, in some way, influence the amount of detail shown. A list of a few things that might be shown includes:

Air sets – Sometimes a symbol is added to pneumatic devices that indicates where instrument air is connected and an air set is needed. An air set is made up of any combination of a pneumatic regulator, a filter and a pressure gauge.

Set points – Some firms add the set points for regulators and switches, although the authors believe that these are better shown on a Loop Diagram.

Root valve – The instrument root valve between the process and the transmitter may have a size and specification called out.

Control valve size – Sometimes the size of the valve is inferred by the size of the piping or by the size of piping reducers; sometimes the size is provided as a superscript outside the instrument bubble.

Valve positioner – In the authors' opinion, the use of a valve positioner can be defined in the construction and purchasing specifications and Installation Details. There is no need to show positioners on the P&ID.

Controller location – The panel, bench board, control room or other location can be added as an identifier outside, but near to, the controller bubble. These will usually appear as an acronym or as a few letters that are further identified on the P&ID legend sheet.

term "Power" means the medium that moves the valve actuator and therefore the valve trim. The most common "Power" medium is instrument air. "Power" does not refer to the signal, unless the signal is the medium that moves the actuator. The loss of power permits the valve spring to move the valve to its fail position

Figure 2-5: Valve Failure

No	Method A	Method B	Definition
1	 FO		• Fail to open position.
2	 FC		• Fail to closed position.
3	 FL		• Fail locked in last position.
4	 FL/DO		• Fail at last position. • Drift open.
5	 FL/DC		• Fail at last position. • Drift closed.

From ANSI/ISA-5.1-2009

Natural Gas Can Substitute for Air

Pneumatic systems did not always use pressurized instrument air as the energy source. Offshore hydrocarbon production platforms had a ready supply of energy in the natural gas produced by the well. For smaller gas production platforms without electric power, a filter dryer served quite well in preparing the pneumatic medium used to provide control and safety systems. Obviously, smoking at work was frowned upon.

The control panels on these facilities were a complex mass of pneumatic tubing, containing specialized components like first-out pneumatic indicators called "winkies."

The fail positions are commonly identified on the P&ID using letters below the valve symbol: FO for Fail Open, FC for Fail Closed, FL for Fail Locked, FL/DO for Fail at Last position, Drift Open, and FL/DC for Fail at Last position, Drift Closed. See Figure 2-5: Valve Failure which shows other methods of indicating the fail positions of control valves. Looking at Figure 2-5, an arrow up signifies that the valve fails open. An arrow down means fail closed. A two-headed arrow is fail locked in last position. Two arrows up is fail in last position, drift open, and two arrows down means fail at last position, drift closed.

It is important to remember that "fail position" refers to the loss of the primary power at the valve, the motive force. Pulling the electronic signal off the valve transducer or electro-pneumatic positioner may induce a different reaction than the failure indication shown above. A springless piston-actuated valve will fail indeterminate upon loss of air pressure. However, if there is a positioner, and no loss of power, the valve will be driven in one direction or the other upon loss of the electronic signal. A valve positioner is an instrument that senses the value of a corrective signal to a control valve and the position of the valve trim and adjusts the value of the power supply to move the trim to the value called for by the corrective signal. Valve positioners are mounted directly on the valve they control.

Figure 2-6 shows an electronic loop controlling flow in a pipeline. The loop number is 101. The dashed line indicates that information is transmitted electronically from the flow transmitter, FT-101, to the electronic indicating controller, FIC-101, and from the controller to the current-to-pneumatic transducer (I/P), FY 101. In this instance, FT-101 senses the differential pressure proportional to the flow rate in the line caused by FE-101, a flow element consisting of an orifice plate. FT-101 transmits a 4–20 mA DC signal corresponding to the varying differential pressure. FIC-101, an electronic flow indicating controller, transmits a 4–20 mA DC signal to the transducer, FY-101, that converts the 4–20 mA signal into a pneumatic signal. This signal changes the position of the valve actuator, which in turn changes the position of the inner works of the control valve, changing the flow rate through the control valve.

Figure 2-6: An Electronic Loop – Controlling Flow

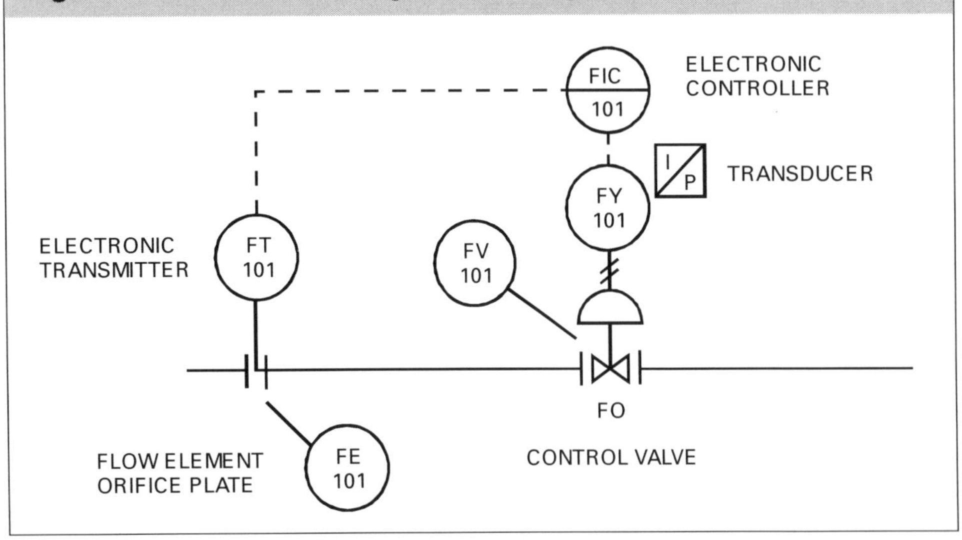

Members of the control systems design group add all the loop and local instruments to the P&ID, one at a time, until the complete instrumentation system is defined on the drawing. The proper location of local instruments should not be neglected, as they can be the first line of contact for those running and maintaining the facility. Your facility can only be improved when the operators and maintenance personnel assist with the P&ID development endeavor.

ISA-5.1

As has been mentioned, ISA-5.1 is the standard most often used in the process industries as the basis for depicting control systems on P&IDs and other documents. It is broad in scope and flexible in use. The following is a quote from ISA-5.1:

"The symbols and identification methods contained in this standard have evolved by the consensus method and are intended for wide application throughout all industries. The symbols and designations are used as conceptualizing aids, as design tools, as teaching devices, and as a concise and specific means of communication in all types and kinds of technical, engineering, procurement, construction, and maintenance documents and not just in Piping and Instrumentation Diagrams."

The basic process control tagging standard for most industrial facilities is based on ISA-5.1. You will find, however, that additional information or interesting interpretations have been added to further define local requirements to meet specific system requirements or even to maintain site tradition.

However they are arrived at and agreed upon, the standards used at your facility must be completely defined and rigidly followed. Without careful control of the symbols and usage, your documentation will rapidly devolve into a mess that will be difficult to understand and use. More important, when the drawings are confusing to read or difficult to work with, people simply stop using them. In addition, drawings and other documentation must be continuously updated to agree with improvements and additions. Without these, mistakes can easily occur. When there is any problem with the use of the drawings, if they are confusing, ambiguous, difficult to read, or inaccessible, they will not be maintained. Drawings that are not maintained with vigilance quickly become useless, or worse, inaccurate.

Device Definition

Each device comprising a control system needs to be uniquely identified for many reasons. The identification string, called a tag or an instrument tag number, serves as a quick way to identify a potentially complex device wher-

To Show or Not to Show?

One of the challenges you will face is the depiction of third party systems on your P&IDs. If you have an island of equipment furnished by a third party, how much of that equipment should show on your drawing? If the third party system suppliers have their own P&IDs, do you copy them into your drawing set, or possibly just include their P&ID with your set? As usual, there really is no right answer; each facility is managed differently, each project has a different scope and each stakeholder in the P&IDs has different requirements.

It is not inexpensive to redraw a P&ID within your drawing set, nor is it a particularly good idea to have two drawings that show the same thing—yours, and the system supplier's. The drawings will probably only agree on the day they are checked and issued for use. As soon as someone makes a change, you start to "chase revisions."

One successful and cost effective approach has been to show the interface points between the vendor's system and your control system—just show the components seen on the operator station. Then, on your drawing, refer to the vendor's P&ID and operating manual for further details.

ever it appears on drawings and in the field. It allows tracking of the device based on what it does, so the tag number can be derived before other identification information, such as the manufacturer and model number, becomes available. The tag is used to track the design, bidding, procurement, shipping

Figure 2-7: Identification Letters

Note: Numbers in paranthesis refer to explanatory notes in ISA-5.1

	First letters		Succeeding letters		
	Column 1	**Column 2**	**Column 3**	**Column 4**	**Column 5**
	Measured/Initiating Variable	**Variable Modifier (10)**	**Readout/Passive Function**	**Output/Active Function**	**Function Modifier**
A	Analysis (2)(3)(4)		Alarm		
B	Burner, Combustion (2)		User's Choice (5)	User's Choice (5)	User's Choice (5)
C	User's Choice (3a)(5)			Control (23a)(23e)	Close (27b)
D	User's Choice (3a)(5)	Difference, Differential, (11a)(12a)			Deviation (28)
E	Voltage (2)		Sensor, Primary Element		
F	Flow, Flow Rate (2)	Ratio (12b)			
G	User's Choice		Glass, Gauge, Viewing Device (16)		
H	Hand (2)				High (27a)(28a)(29)
I	Current (2)		Indicate (17)		
J	Power (2)		Scan (18)		
K	Time, Schedule (2)	Time Rate of Change (12c)(13)		Control Station (24)	
L	Level (2)		Light (19)		Low (27b)(28)(29)
M	User's Choice (3a)(5)				Middle, Intermediate (27c)(28) (29)
N	User's Choice (5)		User's Choice (5)	User's Choice (5)	User's Choice (5)
O	User's Choice (5)		Orifice, Restriction		Open (27a)
P	Pressure (2)		Point (Test Connection)		
Q	Quantity (2)	Integrate, Totalize (11b)	Integrate, Totalize		
R	Radiation (2)		Record (20)		Run
S	Speed, Frequency (2)	Safety(14)		Switch (23b)	Stop
T	Temperature (2)			Transmit	
U	Multivariable (2)(6)		Multifunction (21)	Multifunction (21)	
V	Vibration, Mechanical Analysis (2)(4)(7)			Valve, Damper, Louver (23c)(23e)	
W	Weight, Force (2)		Well, Probe		
X	Unclassified (8)	X-axis (11c)	Accessory Devices (22), Unclassified (8)	Unclassified (8)	Unclassified (8)
Y	Event, State, Presence (2)(9)	Y-axis (11c)		Auxiliary Devices (23d)(25)(26)	
Z	Position, Dimension (2)	(11c), Safety mented System (30)		Driver, Actuator, Unclassified final control element	

From ANSI/ISA-5.1-2009

and installation of the device. It also is the link to calibration, range verification and maintenance records. As can be seen from Figures 2-4 and 2-6, a tag is a combination of identification letters, numbers, and symbols used to define the devices in a loop. The identification letters contain a lot of information as specified in ISA-5.1 and reproduced as Figure 2-7. The number part of the string will be discussed later.

Figure 2-7 consists of twenty-six rows and five columns. The first column lists, alphabetically, twenty-six process variables, or as ISA-5.1 states, the "measured or initiating variable." By starting with a process variable at the left of Figure 2-7 and adding the letters defined in the succeeding columns, the complete function of the control system device is defined. The first letter of any tag name, therefore, will indicate the process variable being measured. The most common process variables in a process plant include:

F – Flow

L – Level

P – Pressure

T – Temperature

There are several letters: C, D, G, M, N and O, which can have a meaning specified by the user. Of course, the user must clearly document the specified meanings on the site P&ID legend sheet, and those meanings should be maintained, without ambiguity or change, for the entire facility or, ideally, the entire company.

Many sites will use ISA-5.1 as the starting point. The legend sheet table can then be modified to incorporate assigned letter designations. Many facilities specifically define acceptable or common letter combinations used at that facility to prevent deviations and ambiguity.

Using X for the first letter is a special case. From ISA-5.1, "First-Letter or Succeeding-Letter for unclassified devices or functions (X), for non-repetitive meanings that are used only once or to a limited extent may have any number of meanings that shall be defined outside tagging bubbles or by a note in the document." The function of the letter is defined on the legend sheet as well as implied with a few descriptive letters adjacent to the bubble. When properly applied the letter X does not appear frequently—only once, or to a limited extent. Instead, the user-defined letters should be used for devices that appear regularly, even if infrequently. Thus, in many modern industrial facilities, X may not be needed, since most devices appear with some regularity. For those of you that have an entire facility filled with XT transmitters or XY transducers, don't worry, this provision of ISA-5.1 is frequently ignored. You are not alone.

Worry only if your documents are inconsistent! The proper use of the X prefix is pointed out here more for those of you who are starting fresh on a new facility.

The second column, marked "Variable Modifier," adds additional information about the first letter, the process variable. For example, if an instrument is used to measure the difference between two pressures, perhaps the upstream and downstream pressure at a filter press, a P for pressure is used as the first letter and a D for differential as a second letter modifier. See Figures 2-8 and 2-9. When instantaneous flow rate is being measured and a totalizer is added to provide total flow over time, the device identification is FQ. The first letter of the tag name is F for flow and the second letter is Q from the second column, signifying integrate or totalize.

The next three columns further define the device. The first of these delineates a readout or passive function. For example, Figure 2-8 shows that the filter press differential pressure PD is measured and indicated, as shown by a third letter G, for gauge. Note that the absence of a dividing line in the middle of the circle (or "bubble") shows that the differential pressure is displayed locally, rather than on a control panel.

Figure 2-8: Filter Press With D/P Gauge

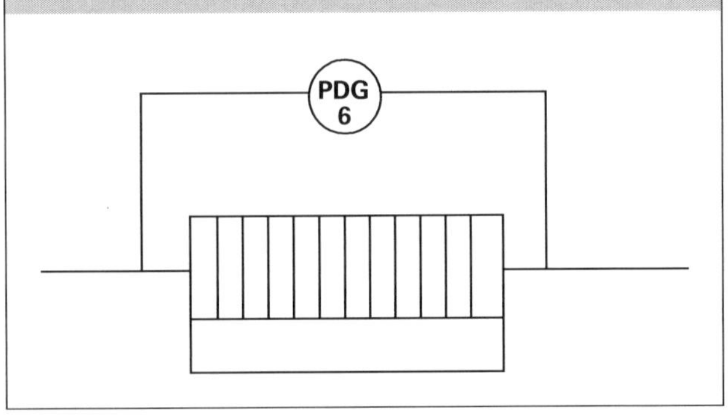

Therefore, PDG shows, locally, the pressure drop across the filter. Figure 2-9 shows that the pressure differential value is transmitted to a central location. The second column of succeeding letters shows that we would use a T for a transmitter, so the device would be a PDT.

Figure 2-9: Filter Press With D/P Transmitter

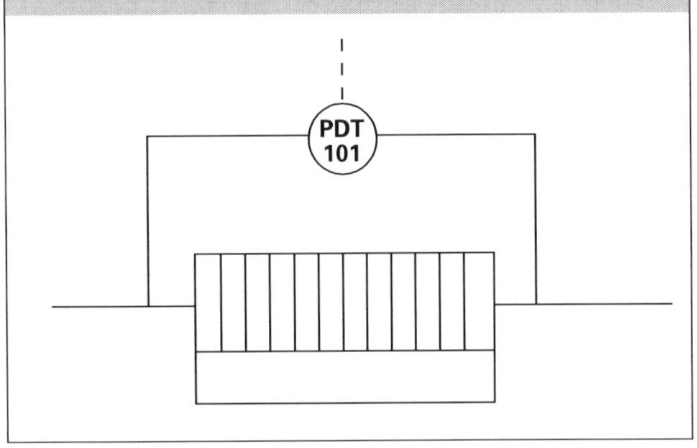

To be certain that we have used allowable function letters for the pressure differential transmitter we can check with Figure 2-10. In the left column we find the first letters PD and in the transmit column we see the letters PDT. This confirms that we have used an allowable combination as shown in Figure 2-9 for the pressure differential transmitter, PDT.

It is good practice to maintain a table of allowable combinations on your facility legend sheet or in your specifications and for your facility to have a procedure for adding new combinations after review and acceptance by some supervising authority. You may be surprised at the number of unnecessary combinations that are put forward by people less attentive to the existence of already approved combinations, particularly when design work is done by multiple design firms.

Figure 2-10: Succeeding Letters

First Letters		B	C				K	N	S		T			U	V	X	Y	Z
Measured/Initiating Variables w/ and w/o Modifiers		User's Choice	Control	Indicate Control	Record Control	Control Valve	Control Station	User's Choice (4a)	Switch	Function Modifier	Transmit	Indicating Transmit	Recording Transmit	Multi-function	Valve Damper Louver	Unclass-ified	Compute, Convert Relay	Actuator, Drive
			C	IC	RC	CV			S		T	IT	RT					
A	Analysis		AC	AIC	ARC	NA	AK		AS[*]		AT	AIT	ART	AU	AV	AX	AY	AZ
AZ	Analysis(SIS))		AZC	AZIC	AZRC	NA	NA		AZS[*]		AST	NA	NA	AZU	AZV	NA	AZY	AZZ
B	Burner, Combustion		BC	BIC	BRC	NA	BK		BS[*]	[*] = Function Modifier	BT	BIT	BRT	BU	BV	BX	BY	BZ
BZ	Burner, Comb.(SIS)		BZC	BZIC	BZRC	NA	NA		BZS[*]		BZT	NA	NA	BZU	BZV	NA	BZY	BZZ
C	User's Choice																	
D	User's Choice																	
E	Voltage		EC	EIC	ERC	NA	EK		ES[*]	High-High HH	ET	EIT	ERT	EU	NA	EX	EY	EZ
EZ	Voltage(SIS)		EZC	EZIC	EZRC	NA	NA		EZS[*]		EZT	NA	NA	EZU	NA	EZX	EZY	EZZ
F	Flow, Flow Rate		FC	FIC	FRC	FCV	FK		FS[*]	High H	FT	FIT	FRT	FU	FV	FX	FY	NA
FF	Flow Ratio		FFC	FFIC	FFRC	NA	FFK		FFS[*]		FFT	FFIT	FFRT	FFU	FFV	FFX	FFY	NA
FQ	Flow Total		FQC	FQIC	FQRC	FQCV	FQK		FQS[*]	Middle M	FQT	FQIT	FQRT	FQU	FQV	FQX	FQY	NA
FS	Flow Safety		NA	NA	NA	FSV	NA		NA	Low L	NA	NA	NA	NA	FSV (10)	NA	NA	NA
FZ	Flow(SIS)		FZC	FZIC	FZRC	NA	NA		FZS[*]		FZT	NA	NA	FZU	FZV	NA	FZY	NA
G	User's Choice									Low-Low LL								
H	Hand		HC	HIC	HRC	HCV	NA		HS		NA	NA	NA	HU	HV	HX	HY	HZ
HZ	Hand(SIS)		HZC	HZIC	HZRC	NA	NA		HZS	Open O	NA	NA	NA	HZU	HZV	NA	HZY	HZZ
I	Current		IC	IIC	IRC	NA	IK		IS[*]		IT	IIT	IRT	IU	NA	IX	IY	IZ
IZ	Current(SIS)		IZC	IZIC	IZRC	NA	NA		IZS[*]	Close C	IZT	NA	NA	IZU	NA	IZX	IZY	IZZ
J	Power		JC	JIC	JRC	NA	JK		JS[*]		JT	JIT	JRT	JU	NA	JX	JY	JZ
JQ	Power Totalize		JQC	JQIC	JQRC	NA	JQK		JQS[*]		JQT	JQIT	JQRT	JQU	NA	JQX	JQY	JQZ
JZ	Power(SIS)		JZC	JZIC	JZRC	NA	NA		JZS[*]	Run R	JZT	NA	NA	JZU	NA	JZX	JZY	JZZ
K	Time, Schedule		KC	KIC	KRC	NA	KK		KS[*]		NA	NA	NA	KU	NA	KX	KY	KZ
KQ	Time Totalize		KQC	KQIC	KQRC	NA	NA		KQS	Stop S	NA	NA	NA	KQU	KQV	KQX	KQY	KZZ
L	Level		LC	LIC	LRC	LCV	LK		LS[*]		LT	LIT	LRT	LU	LV	LX	LY	LZ
LZ	Level(SIS)		LZC	LZIC	LZRC	NA	NA		LZS[*]	Unclassified X	LZT	NA	NA	LZU	LZV	LZX	LZY	LZZ
M	User's Choice																	
N	User's Choice																	
O	User's Choice																	

Note: This table shows allowable succeeding letters for output/active function letters. This table is taken from ISA 5.1, tables A.3.2.1 and A.3.2.2. The numbers in parenthesis refer to explanatory notes associated with those tables in ISA 5.1.

From ANSI/ISA-5.1-2009

Figure 2-10: Succeeding Letters (continued)

First Letters		B	C				K	N	S		T			U	V	X	Y	Z
Measured/Initiating Variables w/ and w/o Modifiers		User's Choice (4a)	Control	Indicate Control	Record Control	Control Valve	Control Station	User's Choice 4a)	Switch	Function Modifier [*] (3)	Transmit	Indicating Transmit	Recording Transmit	Multi-function	Valve Damper Louver	Unclass-ified	Compute, Convert Relay	Actuator, Drive
			C (7)	IC (8)	RC (8)	CV (9)			S	[*] (3) (4d)	T	IT	RT					
P	Pressure		PC	PIC	PRC	PCV (11)	PK		PS[*]		PT	PIT	PRT	PU	PV	PX	PY	PZ
PD	Pressure Differential		PDC	PDIC	PDRC	PDCV	PDK		PDS[*]		PDT	PDIT	PDRT	PDU	PDV	PDX	PDY	PDZ
PF	Pressure Ratio		PFC	PFIC	PFRC	NA	PFK		PFS[*]		NA	NA	NA	PFU	PFV	PFX	PFY	PFZ
PJ	Pressure Scan		NA	NA	NA	NA	NA		PJS[*]		PJT	NA	NA	NA	NA	NA	NA	NA
PK	Pressure Rate of Change		PKC	PKIC	PKRC	NA	PKK		PKS[*]		NA	NA	NA	PKU	PKV	PKX	PKY	PKZ
PS	Pressure Safety		NA	NA	NA	PSV	NA		NA	[*] = Function Modifier	NA	NA	NA	NA	PSV (9)	NA	NA	NA
PZ	Pressure(SIS)		PZC	PZIC	PZRC	NA	NA		PZS[*]		PZT	NA	NA	PZU	PZV	PZX	PZY	PZZ
Q	Quantity		QC	QIC	QRC	NA	QK		QS[*]		QT	QIT	NA	QU	QV	QX	QY	QZ
QQ	Quantity Totalize		QQC	QQIC	QQRC	NA	QQK		QQS[*]	High-High HH	QQT	QQIT	NA	QQU	QQV	QQX	QQY	QQZ
R	Radiation		RC	RIC	RRC	NA	RK		RS[*]		RT	RIT	RRT	RU	RV	RX	RY	RZ
RQ	Radiation Totalize		RQC	RQIC	RQRC	NA	RQK		RQS[*]	High H	RQT	NA	NA	RQU	RQV	RQX	RQY	RQZ
RZ	Radiation(SIS)		RZC	RZIC	RZRC	NA	NA		RZS[*]	Middle	RZT	NA	NA	RZU	RZV	RZX	RZY	RZZ
S	Speed, Frequency		SC	SIC	SRC	SCV	SK		SS[*]	M	ST	SIT	SRT	SU	SV	SX	SY	SZ
SZ	Speed(SIS)		SZC	SZIC	SZRC	SZCV	NA		SZS[*]	Low L	SZT	NA	NA	SZU	SZV	SZX	SZY	SZZ
T	Temperature		TC	TIC	TRC	TCV	TK		TS[*]]		TT	TIT	TRT	TU	TV	TX	TY	TZ
TD	Temperature Differential		TDC	TDIC	TDRC	NA	TDK		TDS[*]	Low-Low LL	TDT	TDIT	TDRT	TDU	TDV	TDX	TDY	TDZ
TF	Temperature Ratio		TFC	TFIC	TFRC	NA	TFK		TFS[*]		NA	NA	NA	TFU	TFV	TFX	TFY	TFZ
TJ	Temperature Scan		NA	NA	NA	NA	NA		TJS[*]	Open	TJT	NA	NA	NA	NA	NA	NA	NA
TK	Temperature Rate of Change		TKC	TKIC	TKRC	NA	TKK		TKS[*]	O	NA	NA	NA	TKU	TKV	TKX	TKY	TKZ
TS	Temperature Safety		NA	NA	NA	TSV	NA		NA	Close	NA	NA	NA	NA	TSV (9)	NA	NA	NA
TZ	Temperature(SIS)		TZC	TZIC	TZRC	NA	NA		TZS[*]	C	TZT	NA	NA	TZU	TZZV	TZX	TZY	TZZ
U	Multivariable		UC	UIC	URC	NA	NA		US[*]		NA	NA	NA	UU	UV	UX	UY	UZ
UJ	Multivariable Scan		NA	NA	NA	NA	NA		NA	Run R	NA	NA	NA	NA	NA	NA	NA	NA
UZ	Multivariable(SIS)		UZC	UZIC	UZRC	NA	NA		UZS[*]	Stop	NA	NA	NA	UZU	UZV	UZX	UZY	UZZ
V	Vibr., Mach. Analysis		VC	VIC	VRC	NA	VK		VS[*]	S	VT	VIT	VRT	VU	NA	VX	VY	NA
VZ	Vibration(SIS)		VZC	VZIC	VZRC	NA	NA		VZS[*]		VZT	NA	NA	VZZU	NA	VZX	VZY	NA
VX	Vibration X-Axis		VXC	VXIC	VXRC	NA	NA		VXS[*]	Unclassified X	VXT	VXIT	VXRT	VXU	NA	VXX	VXY	NA
VY	Vibration Y-Axis		VYC	VYIC	VYRC	NA	NA		VYS[*]		VYT	VYIT	VYRT	VYU	NA	VYX	VYY	NA
VZ	Vibration Z-Axis		VZC	VZIC	VZRC	NA	NA		VZS[*]		VZT	VZIT	VZRT	VZU	NA	VZX	VZY	NA
VZX	Vibration X-Axis(SIS)		VZXC	VZXIC	VZXRC	NA	NA		VZXS[*]		VZXT	NA	NA	VZXU	NA	VZXX	VZXY	NA
VZY	Vibration Y-Axis(SIS)		VZYC	VZYIC	VZYRC	NA	NA		VZYS[*]		VZYT	NA	NA	VZYU	NA	VZYX	VZYY	NA
VZZ	Vibration Z-Axis(SIS)		VZZC	VZZIC	VZZRC	NA	NA		VZZS[*]		VZZT	NA	NA	VZZU	NA	VZZX	VZZY	NA

Note: This table shows allowable succeeding letters for output/active function letters. This table is taken from ISA 5.1, tables A.3.2.1 and A.3.2.2. The numbers in parenthesis refer to explanatory notes associated with those tables in ISA 5.1.

From ANSI/ISA-5.1-2009

As technology advances, some combinations become defunct. For instance, in the past a field mounted flow controller, FC, was quite common. Today, they are rare; instead the controller would be identified with FIC, or with FQIC to better describe the increased functionality available with digital processing. Flow controllers now commonly come standard with local indication and they often totalize as well. Increased functionality is of real interest to people operating the facility, so it is important to note it on the P&ID. Some facilities will add to the acceptable combinations table a note that a specific combination, while valid in the past, is no longer intended to be used for new construction. The note may say something like "Use only with specific approval by the Maintenance Supervisor," which are code words to a designer for "Don't do this."

The allowable letter combinations are shown in ISA-5-1 in separate tables as shown below.

1. Table A.2.1 Allowable letter/number combinations for loop numbering schemes and first letters A to O. P to ZDZ are shown in Tables A.2.2 and A.2.3.

2. Table A.3.1.1 Allowable succeeding letter combinations for readout/ passive functions and first letters A to O. P to ZDZ are shown in Tables A.3.1.2 and A.3.1.3.

3. Table A.3.2.1 Allowable succeeding letters for output/active function letters and first letters A to O. P to ZDZ are shown in Tables A.3.2.2. and A.3.2.3

Table A.3.2.1 is included as Figure 2-10. Reading across Figure 2-10, starting with F for flow rate, the following letter combinations are developed:

FC – Blind flow controller

FIC – Flow indicating controller

FRC – Flow recording controller

FCV – Self-actuated flow control valve; in other words, a regulator

FK – Flow control station, manual loading station with auto-manual switching

FS(*)– Flow switch; replace * with function modifier, (H)igh or (L)ow

FT – Flow transmitter

FIT – Flow indicating transmitter

FRT – Flow recording transmitter

FU – Multi-function device

FV – Flow valve, damper, louver

FX – Unclassified flow device

FY – Compute, convert, relay (I/P)

NA – Not allowable letter combination (FZ)

Interesting interpretations – and an opinion:

An electro-pneumatic transducer, commonly called an I/P, probably has more combinations of "ISA standard" tags than any other control system component. "ISA standard" is used somewhat facetiously, since clearly all approaches cannot be correct, yet you can be sure that someone along the way assured someone else that their particular approach was in accordance with ISA-5.1, Even within a single facility, creative tagging of I/P's may include no tag at all, I/P, IP, FY, XY, NY and so forth. "No tag" can easily occur when the I/P arrives on site pre-mounted to a control valve by the valve vendor, and the control valve is the only tag in your system.

The correct tagging of an I/P is to use the first letter of the loop in which the I/P appears, "the process variable" followed with a Y as the output function "convert". Thus, a flow loop I/P would be an FY. To be crystal clear, the I/P would be written in a function block or a box adjacent to the bubble. The reason for the creative tagging of I/P may be that, with the widespread use of electronic instrument databases, some may see an advantage in developing a unique identifier for an I/P, so a database sort can list all the I/Ps on a project under one identifier independent of the loop it serves.

There will be many I/Ps on a project. The ability to list all occurrences of a component is handy when specifying and purchasing a component. Also, from a practical standpoint this author was once asked "Since they are called I/Ps, why not tag them as I/Ps?" It is hard to argue with that logic! The I/P tag works since there is not another common device that would call for the use of I/P, there isn't a data clash. Detailed explanations to justify the I/P tag start with: "I is the process variable for current, P is pneumatic pressure, so it works." Well, that may be true. It certainly works, but it isn't technically true from an ISA-5.1 view, so some practitioners may be appalled. The process variable letter is intended for the entire loop, not for that one device in the loop, so technically it should be F, P, T, L, etc. The variable P that the loop is measuring or controlling is not listed as an output function. P is pressure only as a process variable, the first letter in a tag string.

Instrument Numbering

In addition to the letters in a tag, the control systems design group assigns a sequence number to each function. All the devices within that function carry the same sequential number—in other words, the loop number. A single loop number is used to identify the devices that accomplish a single specific action—possibly an input and an output signal and components for PID (proportional-integral-derivative) control, an input for remote indication of a process variable, or a manual output. This number, combined with the letter designation, positively and uniquely identifies and links each device within that set, or loop.

The numbers you choose to use might follow the suggestions in ISA-5.1. However, there are many other numbering systems used throughout industry. All the valid numbering systems share one trait: they provide unique identification of each component and they group related devices ("looped" devices) logically.

ISA-5.1 suggests that loop numbering may be parallel or serial. By "parallel," ISA-5.1 means that a process variable letter is coupled with a number to make the unique identifier. Therefore, there may be an FRC-101, a PIC-101, and a TI-101 since F101, P101 and T101 are unique and parallel. Each of those three letter and number sets define a different loop; they may be related, but they are unique. This numbering system can be used effectively when the number is linked to a piece of equipment, like a pump, where all the loops associated with pump 101 would carry that number within the tags as listed above.

A word of caution: Maintaining this link is not as easy as it sounds; for instance, you need to plan how you will number temperature elements if the equipment has more than one measurement point. Another question that might arise: "Is the flowmeter on a pump related to the pump or to some downstream equipment?" Ideally, solutions to these and other situations should be planned out before the numbers are assigned. You might end up making a few passes at numbering before you have a workable approach. Note that this system has the potential advantage of limiting the quantity of numbers used, since you can have multiple loops with one base number.

Serial numbering, as mentioned in ISA-5.1, means using a unique numerical sequence for each loop without the process variable modifier, one number for each loop. Therefore, there may be an FRC-101, an LR-102, a PIC-103, and a TI-104, but not an FRC-101 and an LR-101 since the Flow and Level variables will each get a different number. This is the simplest system to use and it is therefore probably the most common.

A block of numbers is sometimes used to designate certain types of devices. For example, all safety valves might use the 900 series: PSV-900, PSV-901, PSV-902, etc. Obviously, you should use this approach only if you know you will never have more than 100 of that device, and when it offers some significant advantage to your work.

Instrument numbers may also be structured to identify the loop location or service. For example, see Figure 2-11, Instrument Numbering. The first digit of the number may indicate the plant number; hence, FT-102 is an instrument in Plant 1. Another method of identifying the instrument location is with a prefix, for example 2 (area), or 03 (unit), or 004 (plant 4) which identifies the service area of the loop: 2-FT-102 is loop 102 in area 2, or 03-

Figure 2-11: Instrument Numbering

- **Use Basic Number if project is small and there are no area, unit, or plant numbers:**
  - Basic Number FT-2 or FT-02 or FT-002
- **If project has a few areas, units, or plants (9 or less), use the first digit of the plant number as the tag number:**
  - FT-102 (1 = area, unit, or plant number)
- **If project is divided into area, units, or plants:**
  - 1-FT-002
  - 01-FT-002
  - 001-FT-002

FT-102 is loop 102 in unit 03, or 004-FT-102 is loop 102 in plant 4. These numbers can also be combined to show area-unit-plant in one number: 234-FT-102 is a flow transmitter in loop 102, which serves area 2, unit 3 in plant 4. To complete the confusion, remember that the loop number defines the items in the loop, so the loop may serve the area listed above, but a particular device may be physically located in another area.

A variation of this system is to tie the P&ID drawing numbers to a particular area, and then to sequentially number the instruments on that P&ID sheet. For example, P&ID 25 carries up to 100 loops, or instrument loop numbers 2500 to 2599. The elegance of this system is that you can find the correct P&ID for an instrument based upon the tag number alone, since the tag number includes the P&ID number. Frequently, the area number is nested in the P&ID number anyway, so you will also know the area served by the loop just by looking at the loop number.

Many different numbering systems are used. Some incorporate a major equipment number into the instrument identification. Still another variation deviates from the "unique number" requirement by the use of the "loop" number as a coding system to group similar "commodity-type" devices. The number that appears in the loop number place on the instrument circle is a component identifier that is tied to a master device specification. This approach can be useful for calling out stand-alone devices that don't interface to the control system, such as local indicators like pressure gauges.

This type of instrument can be referred to as a commodity-type device. For example, in your facility PG-100 might be listed in the specifications as a liquid filled 4 1/2" diameter pressure gauge with a range of 0-150 psig and a stainless steel Bourdon tube. As long as all your PG-100s are the same, this system works. However, when you have a different material of construction or some other change, a different number has to be used with this system. Of course, a more complete pressure gauge specification can be used when actually purchasing the gauge.

There also might be many component numbers when this system is used on temperature gauges, since there are so many variations of stem length, dial size and range. This approach is therefore not common, but applied with care, it can be useful.

However it is constituted, the numbering system chosen for your P&IDs and loops should be tested and verified to ensure that it works as expected with the various software applications used in your facility. Loop number 1 and loop number 001 will have markedly different treatment by some databases and by the maintenance planning and inventory control software.

Location Information

The letters and numbers that identify loop components have to appear on a drawing somewhere, so the next step is to put the chosen function identifier and loop number, the tag number, on the P&ID. ISA-5.1 provides the information needed to further define the location of an instrument through the use of specific symbols. The symbols are shown in Figure 2-12.

Figure 2-12: Instrumentation Device and Function Symbol

No.	Shared display, Shared control		C	D	Location & accessibility
	A	B			
	Primary Choice or Basic Process Control System	Alternate Choice or Safety Instrumented System	Computer Systems and Software	Discrete	
1	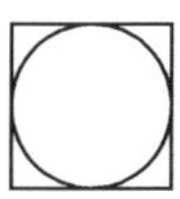	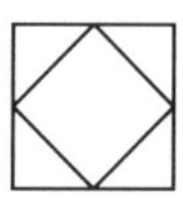		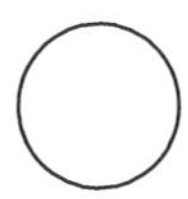	• Located in field. • Not panel, cabinet, or console mounted. • Visible at field location. • Normally operator accessible.
2	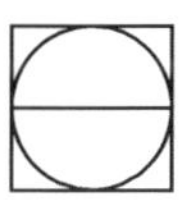	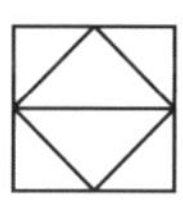	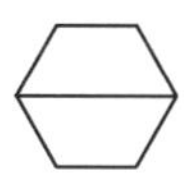	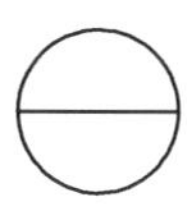	• Located in or on front of central or main panel or console. • Visible on front of panel or on video display. • Normally operator accessible at panel front or console.
3	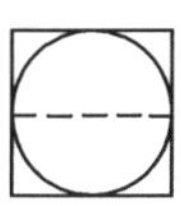	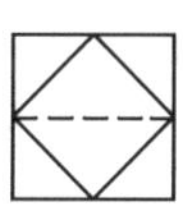	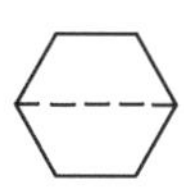	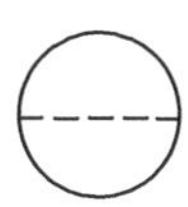	• Located in rear of central or main panel. • Located in cabinet behind panel. • Not visible on front of panel or on video display. • Not normally operator accessible at panel or console.
4	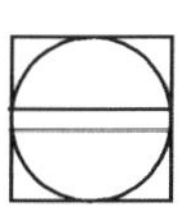	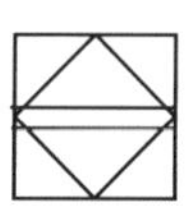	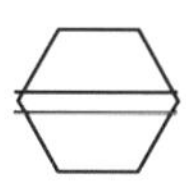	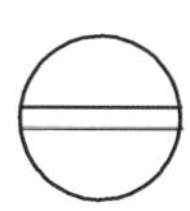	• Located in or on front of secondary or local panel or console. • Visible on front of panel or on video display. • Normally operator accessible at panel front or console.
5	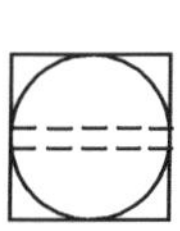	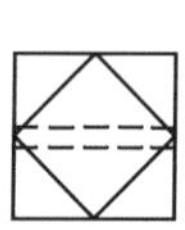	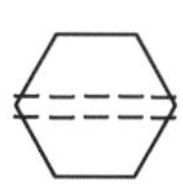	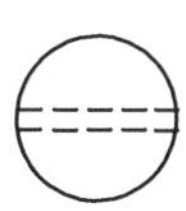	• Located in rear of secondary or local panel. • Located in field cabinet. • Not visible on front of panel or on video display. • Not normally operator accessible at panel or console.

From ANSI/ISA-5.1-2009

The circles, squares, hexagons and diamonds all have meaning. A circle means that the device is field mounted (located in the process area of the plant). If a line is added through the center of the circle, it means that the device is located in the primary location normally accessible to the operator, such as in the central control room or on the main control panel. If a second line is added, parallel to the first, the device is located in an auxiliary location normally accessible to the operator, such as on the face of a starter cassette in a motor control center or, less commonly, on the face of a local panel, assuming that it is called out by your facility standards.

A dashed line through the center of the circle shows that the device is normally inaccessible to the operator, such as inside a panel, which will require someone to open a panel door to access the device. If an external square is added to the circle, the symbols represent devices or functions that are part of a shared-display, shared-control, perhaps a DCS. If we substitute a hexagon for the circle or the squared circle, the symbols represent a computer function. A diamond within a square is used to define functions within an alternate shared-display, shared-control system, perhaps a PLC or a Safety Instrumented System (SIS).

Line Symbols

Figure 2-13, Instrument Line Symbols, is copied from ISA-5.1. Line symbols are used to define the ways information is transferred between the field devices and the central control location or the operator interface point. The symbols describe how signals are transmitted between devices. The lines used should be lighter or thinner than the associated process piping so the piping stands out. Signal lines should be secondary or tertiary in viewing precedence. The process sensing line, the pipe or tubing that connects a pressure transmitter directly to the process, is the lightest acceptable "pipe" line.

A line with a double parallel crosshatch defines pneumatic transmission—usually, but not always, instrument air. (Interestingly, some gas pipelines and offshore platforms use natural gas, while some processes require the use of nitrogen rather than air.) Unguided electromagnetic signals, such as radio, are shown by a series of sine waves. If the sine waves are superimposed on a line, the waves are guided.

Internal system links such as software or data links are shown as a repeated line-and-circle, line-and-circle. This symbol is commonly used for a digital signal. With the increasing use of digital communications, you will see line types developed to show specific protocols by adding letters in place of the circle, such as an E for Ethernet. Again, this is fine as long as your legend sheet identifies the line type.

Figure 2-13: Line Symbols – Part 1

No	Symbol	Application
1	(1) IA ————	• IA may be replaced by PA [plant air], NS [nitrogen}, or GS [any gas supply]. • Indicate supply pressure as required, e.g., PA-70 kPa, NS-150 psig, etc.
2	(1) ES ————	• Instrument electric power supply. • Indicate voltage and type as required, e.g. ES-220 Vac. • ES may be replaced by 24 Vdc, 120 Vac, etc.
3	(1) HS ————	• Instrument hydraulic power supply. • Indicate pressure as required, e.g., HS-70 psig.
4	(2) —/——/—	• Undefined signal. • Use for Process Flow Diagrams. • Use for discussions or diagrams where type of signal is not of concern.
5	(2) —//——//—	• Pneumatic signal, continuously variable or binary.
6	(2) - - - - - - - - -	• Electronic or electrical continuously variable or binary signal. • Functional diagram binary signal.
7	(2) ————	• Functional diagram continuously variable signal. • Electrical schematic ladder diagram signal and power rails.
8	(2) —L——L—	• Hydraulic signal.
9	(2) —X——X—	• Filled thermal element capillary tube. • Filled sensing line between pressure seal and instrument.
10	(2) —∿——∿—	• Guided electromagnetic signal. • Guided sonic signal. • Fiber optic cable.
11	(3) a) ∿ ∿ b) ↯ ↯	• Unguided electromagnetic signals, light, radiation, radio, sound, wireless, etc. • Wireless instrumentation signal. • Wireless communication link.
12	(4) —o——o—	• Communication link and system bus, between devices and functions of a shared display, shared control system. • DCS, PLC, or PC communication link and system bus.
13	(5) —●——●—	• Communication link or bus connecting two or more independent microprocessor or computer-based systems. • DCS-to-DCS, DCS-to-PLC, PLC-to-PC, DCS-to-Fieldbus, etc. connections.
14	(6) —◇——◇—	• Communication link and system bus, between devices and functions of a fieldbus system. • Link from and to "intelligent" devices.
15	(7) --o·----·o--	• Communication link between a device and a remote calibration adjustment device or system. • Link from and to "smart" devices.

From ANSI/ISA-5.1-2009

Figure 2-13: Line Symbols – Part 2

	Symbol	Description
16		• Mechanical link or connection.
17	(3) a) a) b) b) 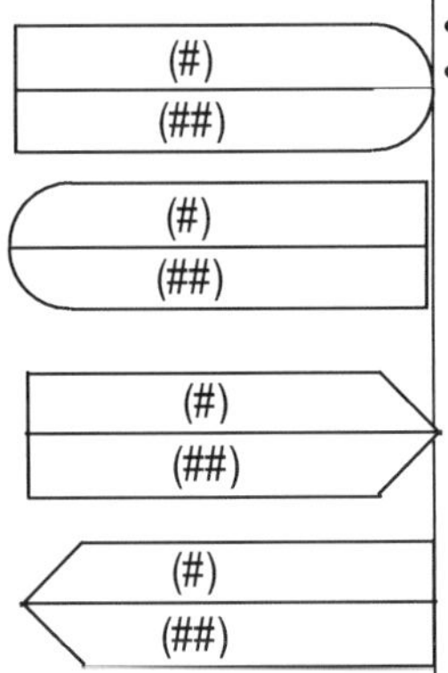	• Drawing-to-drawing signal connector, signal flow from left to right. • (#) = Instrument tag number sending or receiving signal. • (##) = Drawing or sheet number receiving or sending signal.
18		• Signal input to logic diagram. • (*) = Input description, source, or instrument tag number.
19		• Signal output from logic diagram. • (*) = Output description, destination, or instrument tag number.
20	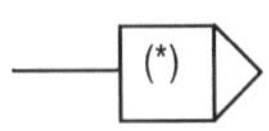	• Internal functional, logic, or ladder diagram signal connector. • Signal source to one or more signal receivers. • (*) = Connection identifier A, B, C, etc.
21	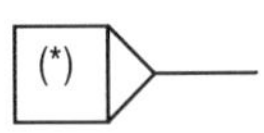	• Internal functional, logic, or ladder diagram signal connector. • Signal receiver, one or more from a single source. • (*) = Connection identifier A, B, C, etc.

From ANSI/ISA-5.1-2009

Pneumatic Transmission

A complete pneumatic transmission system is shown in Figure 2-14. For the purposes of this example, pneumatic signal pressures are 3–15 psig. As has been mentioned, signal pressures can also be 6–30 psig, albeit less commonly. PT-6, a field mounted pressure transmitter, develops and transmits a 3–15 psig signal proportional to the pressure at that point in the process. The signal is transmitted to a field-mounted indicating controller, PIC-6. The controller develops and transmits the 3–15 psig corrective signal to control valve PV-6.

If the valve operator (actuator or "top works") can move the control valve through its entire range with the 3–15 psig signal, regardless of the process pressure, the pneumatic line will be connected directly to the valve operator. If the 3–15 psig signal is not sufficient to operate the valve for all of its design conditions and range, a positioner is added to the valve operator. The function of the

Figure 2-14: Pneumatic Transmission

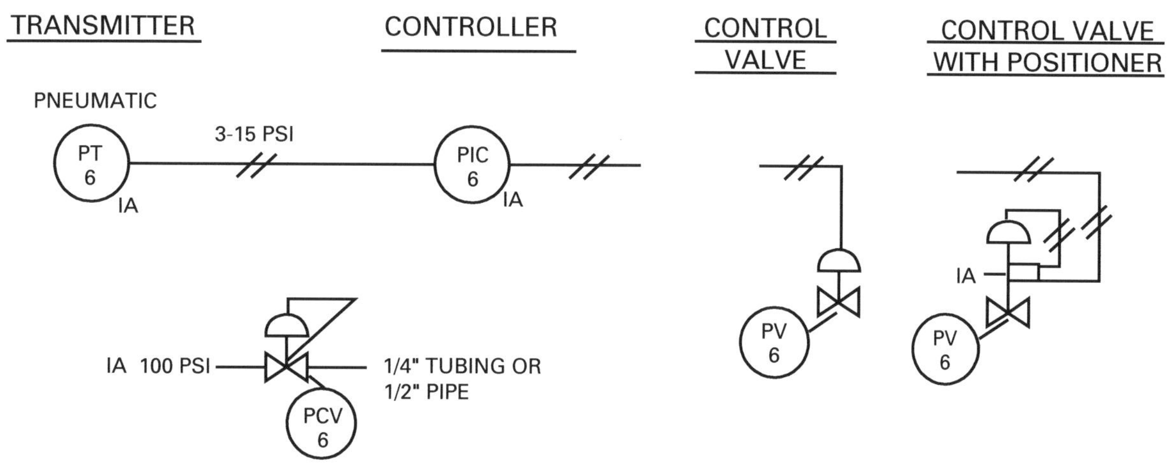

positioner is to compare the incoming signal with the actual valve position and develop the output air pressure necessary to position the valve in accordance with the incoming signal. The output pressure from the positioner to the valve is at a higher pressure, normally 30 psig up to and including full instrument air line pressure of 100 psig or higher.

Figure 2-14 shows that we need a source of instrument air (IA) at the transmitter, another at the controller, and another at the valve positioner. The IA is usually distributed in the field by a complete instrument air piping system, often at a nominal line pressure of 100 psig. Pressure regulators, shown in the figure as PCV-6, are located at the individual devices to reduce the instrument air pressure to that required by the field devices. Pressure regulators that serve pneumatic devices do not always, or—depending upon your industry—do not commonly carry loop identifiers. They may appear on the drawings as an untagged symbol, like a darkened triangle or some variation of an "A" with a line through it to the pneumatic device served.

Electronic Transmission

Figure 2-15 shows a typical electronic transmission system.

The electronic signal transmission system for control systems is often called "a two wire system," meaning that the field transmitter has only two wires connected to it. The signal transmitted usually is nominally 24 v DC with a range of 4 mA to 20 mA, although some installations use 1–5 v DC or, albeit rarely, 10–50 mA DC. Most control valves are pneumatically operated, so even in a modern electronic control system the electronic signal will be converted to a pneumatic signal, which is used to change the position of the valve. The

Figure 2-15: Typical Electronic Transmission

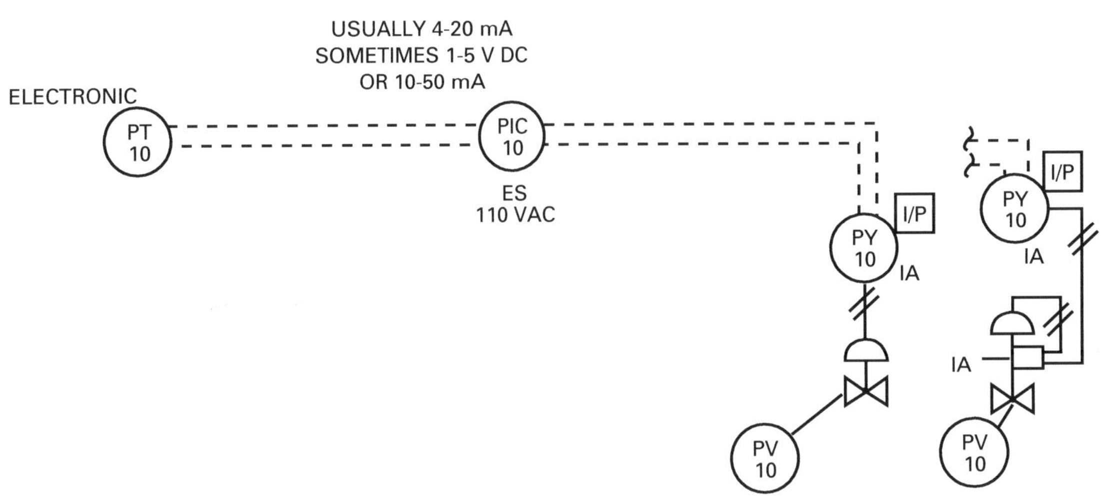

device that does this is a signal converter commonly called an "I to P," written as I/P, which is the abbreviation for a current (I) to pressure (P) transducer. Sometimes the signal conversion is integral to the valve positioner; this assembly is called an electro-pneumatic positioner. The I/P is tagged on drawings and in specifications as a PY if the loop is a pressure loop; see Figure 2-15. The PY tag is the traditional tagging convention. P is for pressure and Y is for "solenoid, relay or computing device" in accordance with the tag letter identification table in ISA-5.1. See Figure 2-7: Identification Letters.

To clarify further, a "function block," a small (1/4") square surrounding the letters I/P, can be added to the right of the converter instrument circle.

A pneumatic or electro-pneumatic positioner is frequently not tagged separately from the valve. It may be left off the drawings because the positioner is usually supplied integral with the control valve, rather than shipped separately. Consequently, there is little need to track it independent of the valve. However, for those times it's needed, there is a symbol and a tag for positioners (ZC) included in ISA-5.1. Symbolically, a simple box on the stem of a control valve can be used to indicate the presence of a positioner. An electro-pneumatic positioner is indicated when the electronic signal terminates on the box instead of a pneumatic signal.

There are many other symbols included in ISA-5.1 for specific instruments. We will not try to show them all.

Final Control Element Symbols

The general valve symbol, the "bow tie," may be used to indicate the body of a control valve or a hand-operated valve. Some projects use this symbol as a generic control valve symbol rather than trying to define the control valve type by using the butterfly or ball symbols in Figure 2-16.

Additional valve symbols are shown in Figure 2-16. The symbol for safety or relief valves consists of an angle valve combined with a spring. Pressure regulators (PCVs) are control valves with actuators, but without an external control signal—designated in the example as either a back pressure regulator or a pressure reducing regulator. The pressure sensing line is shown upstream, if the PCV controls back pressure, and downstream if it controls downstream pressure.

> **Common Misconception**
>
> According to ISA-5.1, it is not correct to use the succeeding letters CV for anything other than a self-actuated control valve. A control valve in a flow loop is identified as an FV. FCV is a self contained flow regulator.

Figure 2-16: Valve Symbols

Symbol	Description
a) b) 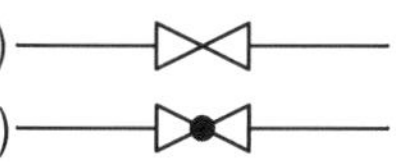	• Generic two-way valve. • Straight globe valve. • Gate valve.
	• Butterfly valve.
	• Ball valve.
	• Backpressure regulator. • Internal pressure tap.
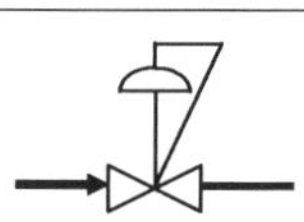	• Pressure-reducing regulator. • Internal pressure tap.
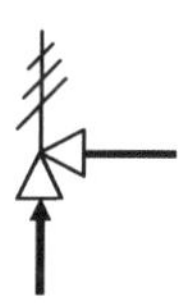	• Generic pressure safety valve. • Pressure relief valve.

From ANSI/ISA-5.1-2009

Primary Flow Measurement Element Symbols

One of the most common means of measuring volumetric flow rate (commonly abbreviated as simply "flow") and transmitting that measurement is with a differential pressure (d/p) cell and orifice plate, one of many types of "primary flow elements."

Figure 2-17: Orifice Plate Primary Elements

A.1.1 Orifice plate primary elements, with or without optional flow arrow, use generic orifice plate symbol with transmitter bubble connected to indicate orifice tap location for flange taps, corner taps, pipe taps, and vena contracta taps respectively:

a) Single process connection: corner taps, pipe taps, and vena contracta taps are indicated by notation:

b) Double process connection, pipe taps and vena contracta taps are indicated by notation:

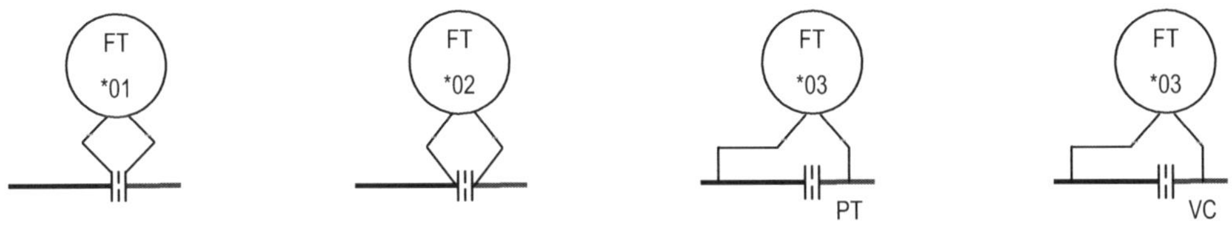

From ANSI/ISA-5.1-2009

Figure 2-18: Measurement Symbols: Primary Elements

Symbol	Description
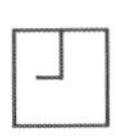	• Standard pitot tube.
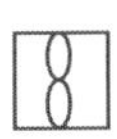	• Turbine flowmeter. • Propeller flowmeter.
	• Vortex shedding flowmeter
a)  b) 	• Magnetic flowmeter.
	• Positive displacement flowmeter
	• Variable area flowmeter

From ANSI/ISA-5.1-2009

Figure 2-17 shows several variations of primary flow elements that produce a pressure differential relative to flow: an orifice plate and flanges with flange taps, corner taps, pipe taps and vena contracta taps.

Other means of flow measurement are shown in Figure 2-18: a pitot tube determines flow by measuring flow pressure against the tube orifice; a turbine meter measures the varying rotational speed of a turbine in a flow stream; a variable area meter, also known as a rotameter, measures flow through the relative position of a "float" or plummet in a graduated tube; a positive displacement device, such as the water meter at your house, is used to measure liquid flow rate. A magnetic flowmeter measures the very small voltage developed when a conductive liquid passes through a magnetic field. A vortex shedding meter measures the change in flow rate in a process stream as a vortex develops and recedes.

Exercise

We have presented an overview of the symbols in ISA-5.1. As a review, please do the following exercises.

Figure 2-19: Descriptions

Instructions: Match the drawing/symbols on the next page with the instrument function title/description below.

1. ( ) Pneumatic Signal, Continuous, Variable, or Binary
2. ( ) Discrete Instrument - Located in Main Panel or Console
3. ( ) Safety Valve
4. ( ) Discrete Instrument - In Secondary Panel or Local Panel or Console
5. ( ) Board Mounted Electronic Level Indicating Controller
6. ( ) Butterfly Valve
7. ( ) Back Pressure Regulator - Internal Pressure Tap
8. ( ) DCS, PLC or PC Communication Link and System Bus
9. ( ) Discrete Instrument, Located in Rear of Main Panel
10. ( ) Shared Display or Control - Basic Process Control System, Main Panel or Console
11. ( ) Unguided Electromagnetic Signal
12. ( ) Electric or Electronic Signal
13. ( ) Variable Area Meter (Rotameter)
14. ( ) Generic Two-way Valve & Operator, Fail Open
15. ( ) Undefined Signal
16. ( ) Discrete Instrument - Field Mounted
17. ( ) Generic Two-way Valve & Operator, Fail Closed
18. ( ) Sight Glass
19. ( ) Pressure Gage
20. ( ) Safety Instrumented System on Main Panel

Match the descriptions from Figure 2-19 with instrument symbols taken from Figure 2-20. When you have finished, check your answers with the answer sheet in Appendix A.

PFD Defines Process Conditions

As the design progresses, information from the PFD is used to define process conditions for equipment and piping. The equipment or vessel designer sizes vessels using information first established on the PFD. The piping designer, or perhaps the process designer, calculates the pipe sizes. The mechanical equipment designer selects equipment. Equipment requirements may influence the process throughput, which has an iterative impact on the process design. Equipment changes introduce process changes that change line sizes. Equilibrium is reached eventually, as the project team establishes more details and pertinent information becomes available.

Figure 2-20: Instrument Symbols

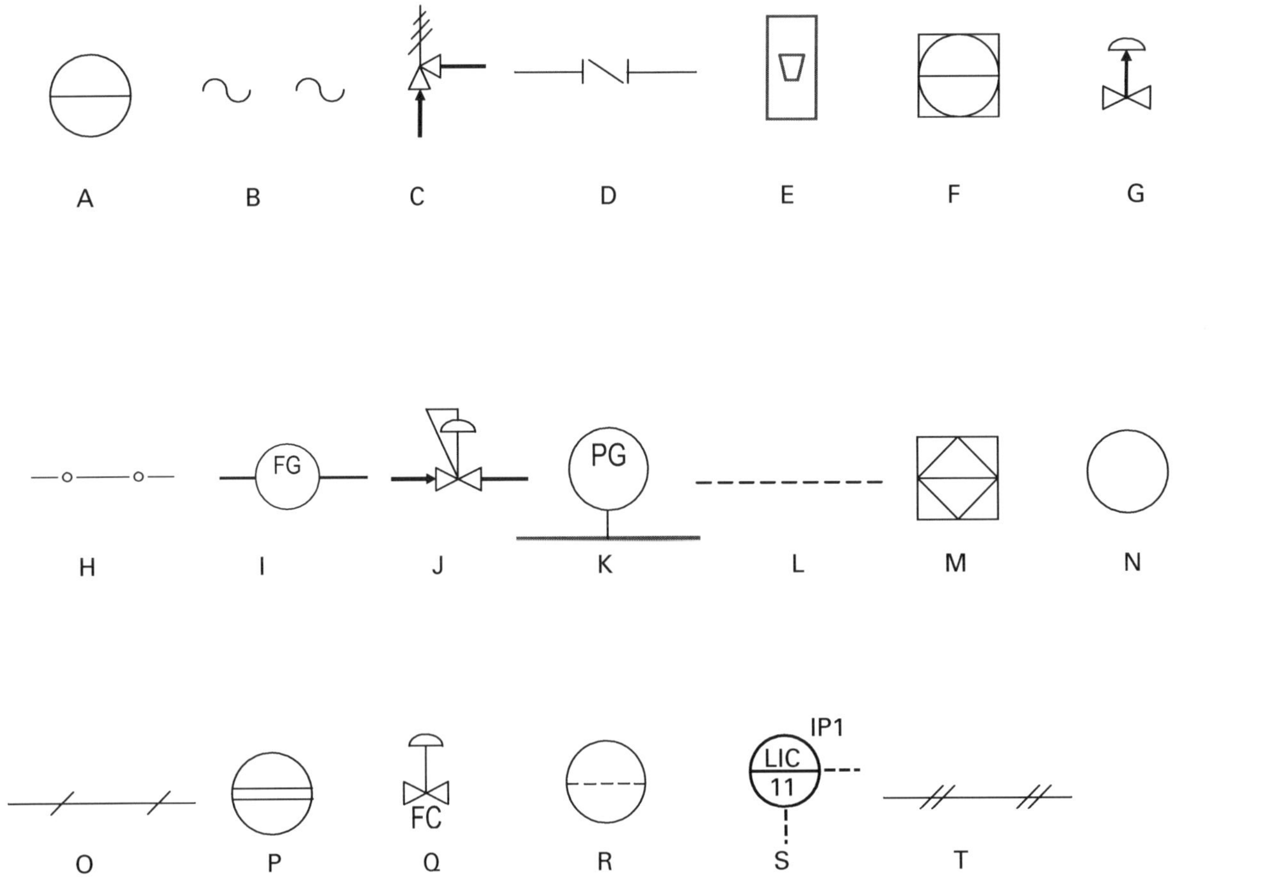

All this information is recorded and updated on the P&ID. The P&ID is the coordinating document among all the designers. Each designer will continually contribute information to the P&ID and verify and use the information added by other groups. As piping and equipment details become available, the control systems designers establish the process sensing points, calculate the control valve sizes and begin to add control loop definition.

Information on a P&ID is frequently a coded reference to more complete data that is maintained separately, such as on Specification Forms (data sheets) and in databases. The number shown on a pipe might be an alphanumeric coded callout. The callout is a link to more detailed information on a line list and in the piping specification. This additional information might include materials of construction, pressure ratings, connection methods and service.

Detailed Design

At some point, the decision-makers for a design effort will conclude that the P&IDs are sufficiently developed to commence with detailed design. The overall effort now transitions into high gear. Some projects mark this point as a schedule milestone. The P&IDs, and perhaps other drawings and documents, are formally issued for detailed design to kick off this effort. The drawings and other documents are identified on the revision blocks that they are "issued for detailed design," or some similar wording. This point corresponds to a significant ramp-up in effort, therefore staffing will likely increase in support of the generation of detailed design documents.

The process control design group then increases its activity to place symbols and tag numbers on the P&IDs to depict each control system device and function. Instrument tag numbers indicate the process variable and the device's function. The tag number provides the knowledgeable user with a link to more complete information on the Specification Forms and other control system documents.

There are no universal standards that address the format to be used in developing P&IDs. The format used by most design groups has been developed over many years. However, here are a few guidelines that serve as a simple *de facto* standard:

- The process flows on the drawing from left to right and, if practical, from top to bottom.
- P&IDs are developed as "D" size sheets (22" x 34") or larger, but should remain legible when reduced to "B" size (11" x 17") for ease of use in the office and in the field.
- P&IDs should show sufficient information to define the process, but without crowding. One to three pieces of equipment with auxiliaries are frequently sufficient for one drawing.
- To reduce clutter, a typical detail can be used for repeated components (see the "typical drain" detail on Figure 2-21).
- When piping gets complex, auxiliary P&IDs can be used.
- Add notes for understanding and clarity.
- It is helpful to show the relative sizes of equipment, but do not include specific elevations or dimensions.
- Every set of P&IDs should include a legend sheet, or sheets, to define the symbols and abbreviations used. Legend sheets should not change as different designer's service your facility; the drawings should be as similar as possible.
- The free space on a P&ID should facilitate the addition of future process changes; it is best not to start with congested P&IDs.

There is always a trade-off when you are deciding how much information and detail should be included on P&IDs. Some information is useful to only one design discipline. Specialists tend to want more of their information presented;

operations staff may want to show a minimum of information to keep P&IDs uncluttered. Therefore, it is always good to think through your standards by answering the questions, "Is this information really of value to the end users of the drawing?" and, "Is this information displayed and maintained elsewhere?" If the information presented is shown somewhere else, you should think carefully about its contribution to the P&IDs and those who use them.

It is common for a designer to think certain information is important, and for an owner or operator to not see the point. Part of setting up P&IDs and other documents must be to decide whose opinion will prevail. One would hope that the simple approach is the successful one; complex justifications rarely survive the departure of a zealous champion.

It is important to always balance the importance of the information you plan to show on a drawing with the expense of maintaining that information. In deciding to depict the actual control valve type through the use of a specific valve symbol, you should ask if that information is germane to the drawing: will the information be of significant use to your team and to others who will make use of the drawing in the future?

For example, does anyone need to know that a butterfly valve was installed, to understand the process? The cost of the information is the expense of maintaining the correct symbol. The control valve symbol is one example of the questions that should be asked when deciding what goes on a P&ID and what does not.

While it is true that P&IDs evolve during a design project, so many changes will be made to the drawings that the actual selection of control valve style might not be known until the valve is specified and purchased, which should be long after the P&IDs have been issued for detailed design. You may be pretty sure a valve will be a globe valve, but you really won't know until the valve is purchased. If you are showing the actual valve type on the P&ID, someone will have to review each valve symbol after the control valves are purchased to ensure that the correct valve type symbol was chosen. There is a cost for that review and correction, and, even more so, for re-issuing the drawings. On a large project, the cost of copying and distributing the drawings can be astonishingly high. Furthermore, once the P&ID has been issued, the devices have been purchased and installed, and the project is complete, the details regarding a particular device are available elsewhere—on Loop Diagrams, Data Sheets, the Instrument Index, etc.

In this book, the project team will use the Process Flow Diagram from Chapter 1, Figure 1-1, to develop a P&ID, Figure 2-21. The P&ID includes KO Drum 01-D-001 and its associated equipment, piping and measurement and control components.

Figure 2-21: P&ID

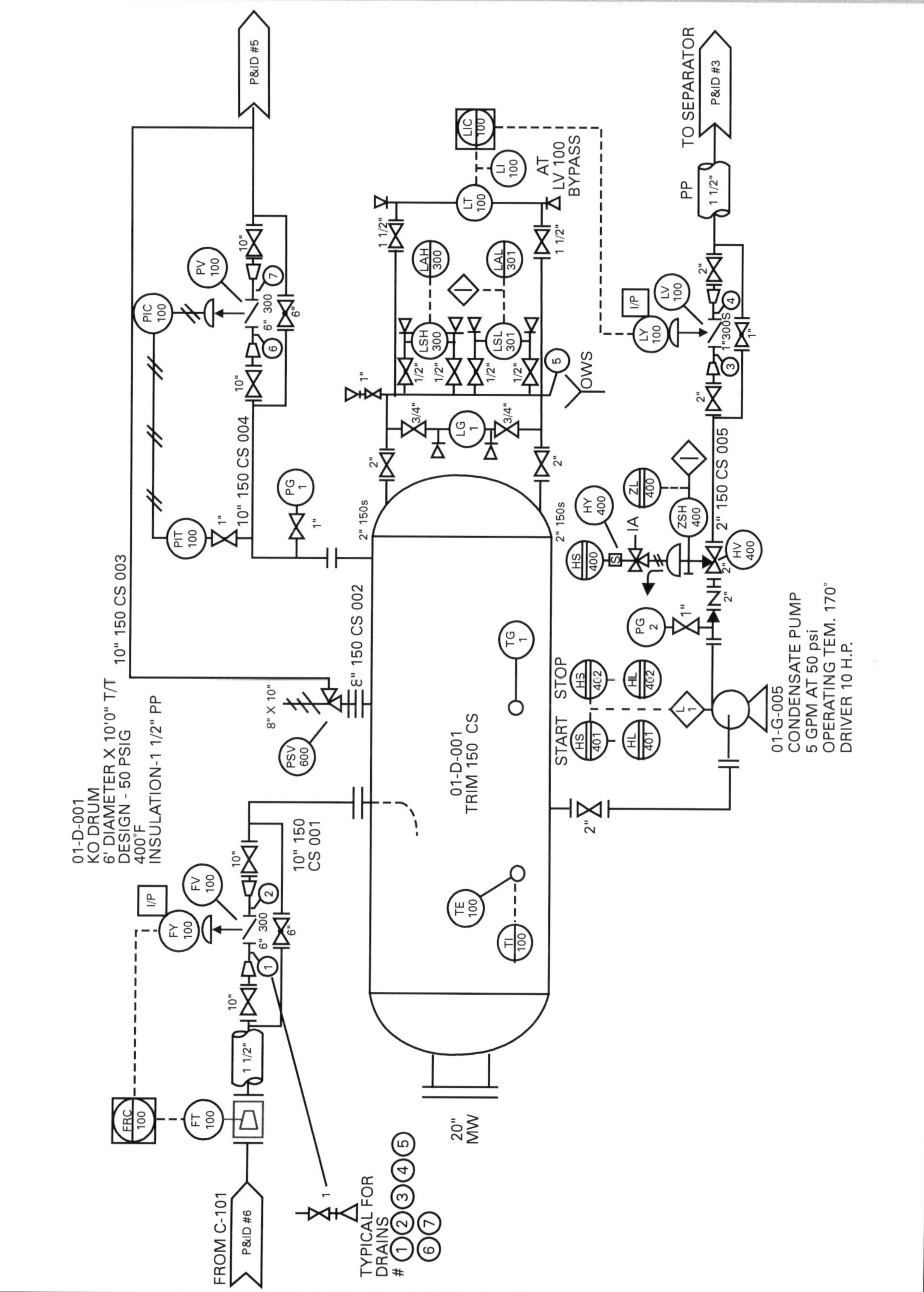

Figure 2-21 includes examples of several control loops including several different methods of documenting the control system. It shows how information might be displayed on a P&ID, but it is not meant to be a realistic design for a KO Drum and its associated equipment.

Note: The drawings developed during design will be identified, at minimum, by a drawing number and a revision number or letter. This information and more is included in a title block. However, for simplicity and to conserve space, title blocks have not been shown on the P&IDs included in this chapter. We will address drawing numbers, revision numbers and letters, and title blocks in Chapter 9.

Equipment Identification

A unique number is normally used to identify equipment on P&IDs and elsewhere in the documentation and Maintenance Management systems. The identifier shown on the P&ID should match exactly the identifier used in, for example, the Maintenance Management system, the specifications, the on-line Operations and Maintenance manuals and so forth. Hopefully you can see the value in having a short, unique identifier to use in searches to provide accurate specification and Maintenance information.

Looking at equipment number 01-D-001, it appears in at least two places—on the PFD and on the P&ID. But wait, there's more! The D-001 used on the PFD has now been expanded to include prefix 01, to signify that the drum is located in plant 01. In this example, the rating for the trim piping, ANSI 150 carbon steel, is shown directly on the vessel symbol, 150CS. (Trim piping is the piping necessary to connect instrument and vessels.) Additional information is frequently shown at the top and the bottom of the P&ID drawing. In particular, equipment rating, size and nominal or design throughput are shown. However, there is no standard way of depicting this information. Some drawings show equipment information within or adjacent to the symbol itself. Others may show the detailed information along the top or bottom of the sheet, relying on proximity and the reader's knowledge to connect the symbol and the data. As an alternative, the equipment number may be used to link the data and the symbol.

The vessel designers have specified vessel 01-D-001 and the P&ID symbol reflects that design. It is a horizontal vessel, six feet in diameter and ten feet from tangent to tangent, a common measurement for a vessel. The tangent lines define the cylindrical part of a vessel; the head design completes the shell. Also called out are a 20" diameter access port (formerly referred to by knuckle-dragging old timers as a manway, MW) and the internal piping to direct the incoming wet gas.

As shown with the data text on the drawing, the vessel is designed to withstand a maximum internal pressure of 50 psig at 400°F. The two simple letters PP within the vessel information text convey a lot of information; "1 ½" PP" means the vessel will be insulated with 1½" of insulation for personnel protection. This insulation is placed on equipment and piping to protect personnel from injury through contact with a hot surface. The entire vessel is not insulated because we want the gas to cool. The portion of 01-D-001 to be insulated is defined in a specification—perhaps just the area that can be reached from the ground or a platform, or the areas that might be touched when climbing a permanently installed ladder. (Before you ask, personnel protection for someone using a portable ladder will likely be addressed by a JHA – Job Hazard Analysis for that task).

Our P&ID shows pump 01-G-005 information adjacent to the pump symbol. Other designs may show more or less information elsewhere. The designers have elected not to show a symbol for the pump driver, although some designers may include this information. They elected to not show the driver because there is only one type of driver at this facility, an electric motor. The driver, in this case, is understood and showing it is unnecessary clutter. However, pump driver start and stop information is shown because operators want to know this for training and because designers may need it in the future to refresh their memories. If almost every motor in the facility will be started and stopped the same way, which is theoretically possible, the pump driver start and stop information could be provided by a single typical symbol, with details furnished for exceptions only.

Piping and instrument process connections may now be sized and shown on the P&ID. Sizing and showing these connections requires coordination among several groups.

For example, the process control group has added the symbols for a thermocouple (TE-100) and its thermowell (the small circle) to the P&ID. See Chapter 8, Figure 8-1. The vessels group and the process control group agree, after consulting the project and plant specifications, that the vessel connection, size and type will be a 1" 3,000 psig threaded coupling. The agreed-upon locations will be included on the vessel design drawing and on the Location Plan. (See Chapter 8 for more information on Location Plans.)

The vessel design drawing is typically a stylized plan and elevation schematic showing the layout of the vessel with a connection schedule listing the purpose, size, rating, connection information and location of all the connections furnished by the vessel fabricator. Remember that the vessels go out for bid, purchase and fabrication relatively early in the project, so the control system designers must focus on defining their connection requirements early to support that schedule.

The piping designers determine the size and rating for the main process piping, based on the information on the PFD and the specifications. They add this information as well as line numbers to the P&ID and then flesh out the balance of the secondary piping information as the design progresses. The vessel group uses this information to specify the vessel connection size and type.

Our P&ID uses a simple line numbering system to identify all piping lines. There is no industry standard for line numbers, although at the time of this writing, PIP and others may be addressing the issue. Our system includes the line size in inches, the pressure rating of the line in pounds per square inch, the material of the line by an abbreviation, and a sequence number. For example, the incoming line to vessel 01-D-001 is 10" 150 CS 001. This is a 10" diameter carbon steel line rated at ANSI Class 150 and identified by sequence number 001.

Since there is no industry standard for piping line numbers, other designers might show more or less information than that on a P&ID. Some designers show only a sequential number on the P&ID, with all additional information shown in a separate line list or in a separate database. Some line lists show information defining the start and end of the line: "From" and "To" information. The pipe schedule (wall thickness) and nominal pressure rating are often provided on the P&ID. Design flow rates can be shown on the line list, but this information might be better used, maintained and coordinated using the P&ID.

Service Designation Abbreviations

Other designers may include more complex line numbers on the P&ID. Many include a symbol or an abbreviation for the service, which calls out the material flowing in the line. A sampling of these abbreviations includes:

A - Air	FO - Fuel Oil	S-25 - 25 psig Steam
C - Condensate	IA - Instrument Air	S-100 - 100 psig Steam
CW - Cooling Water	N - Nitrogen	PA - Plant Air
FG - Fuel Gas	S - Steam	PW - Potable Water

As with symbols, it is important to diligently control the initiation and use of these abbreviations. To one designer PW might be potable water, but to another it could be plant water, which is not suitable for drinking. Too precise a definition can also be a problem. You have to decide when cooling water is CW and when it is PW; there ought to be a significant and easily understood reason for the different callout.

The line list or database might include additional information about the flowing material—for example, flow rate in gallons per minute, pounds per hour, or cubic feet per minute; pressure, temperature, viscosity, density and specific gravity. Agreed-upon units of measurement should be used throughout the P&ID set. For instance, liquid flow rate may always be given in gallons per minute, steam in pounds per hour, air in standard cubic feet per minute, and gases in units typical for your industry. The paper industry may use units of bone dry tons of wood fiber per day. Bone Dry is defined in the *ISA Dictionary* as a paper making term used to describe pulp fibers or paper from which all water has been removed, also known as "oven dry" and "moisture free".

The units are probably metric if the parent company is European or Canadian. To minimize the space needed for line identification, units are frequently not listed on the line callouts; they are only shown on the legend sheet. No matter what your units are, you need to define them on the legend sheet early in the design and apply them consistently. It may not be rocket science, but poor coordination of units can cause spacecraft to crash, as NASA has learned.

Concurrent with the foregoing P&ID development, the control system designers must decide on the overall control scheme. This decision usually occurs after consultation with the operators and the process designers. Both of these groups have valuable information on the best means of control for a process.

On our example project, the control system designer has added three control loops to the P&ID: flow loop F-100, pressure loop P-100 and level loop L-100.

First, there is a flow loop (F-100) on line number 10" 150 CS 001. In Figure 2-22, the electronic flow transmitter (FT-100) uses vortex shedding technology. It measures flow rate by the changes in the downstream vortex of the flowing fluid. The vortex is introduced into the fluid by a bar placed across the flowing stream. The controller (FIC-100) is part of a shared display control system—in our example facility a DCS located in the central control room or, as stated in ISA-5.1, "Located in or on front of central or main panel or console."

The instrument signals are transmitted electronically, as shown by the dashed lines. The electronic output of the controller, the control output, is converted to a pneumatic signal by the I/P (current-to-pneumatic) converter, FY-100, installed on the control valve, FV-100. FV-100 is a butterfly control valve with a diaphragm oper-

Figure 2-22: Flow Loop F-100

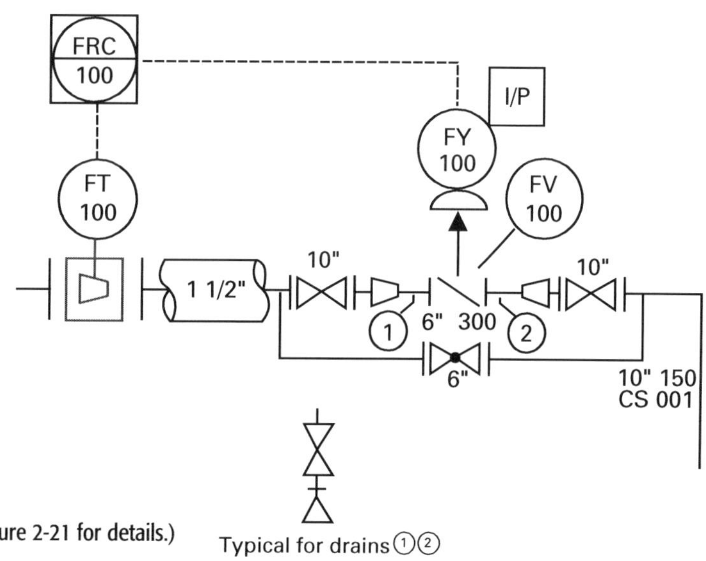

(Refer to Figure 2-21 for details.)

ator which includes a spring to open the valve. If the air supply to the valve is lost, the valve will fail to the open position, as shown by the upward arrow on the valve operator. FV-100 is installed with block and bypass valves, as defined, hopefully, in the owner's design requirements specification. Still referring to Figure 2-22, note the valves on either side of each to the control valves and the valve parallel to the control valve. These are called block and bypass valves. The block or isolation valves permit the control valve to be isolated for maintenance while limiting the amount of adjacent piping that has to be depressurized, drained and cleaned for access. It is possible, sometimes, for experienced staff to physically manipulate the hand operated bypass valve as a rudimentary control method to allow the process to remain in operation should the control valve fail: "It may not be pretty, but it works."

The control system design group has calculated the size of the control valve using the flow, pressure drop and temperature agreed upon with the piping and process designers. The control valve size is 6" and its flanges are rated at ANSI 300 in accordance with the specifications. The piping designers, therefore, show reducers in the line to and from FV-100 while using full-line-size (10") shut-off valves. Full size shutoff valves are used in this case to ensure that the greatest amount of pressure drop is available for the control valve's use. The size of the block valves adjacent to a control valve should be factored into the control valve sizing. Generally, the smaller a valve, the greater the flow restriction and therefore the greater the pressure drop. The ratio of the size of the piping adjacent to a control valve and the size of the valve itself also has an impact on the control valve size calculation, but that is the subject of other books.

The control system designers size and document all the relevant control valve information on Specification Forms or data sheets. They send this information to Purchasing to acquire the valves. The selected valve manufacturer supplies dimensional information for each valve to the designers, including the piping group, so they may complete the detail piping drawing.

Figure 2-23: Pressure Loop P-100

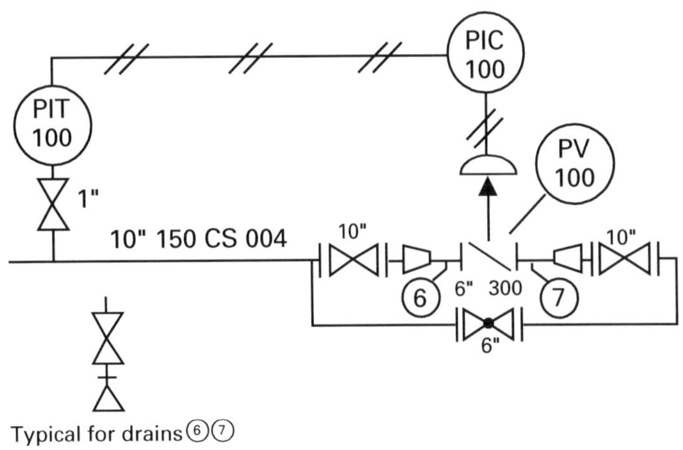

(Refer to Figure 2-21 for details.)

Second, there is a pressure loop, P-100, on line 10" 150 CS 004 (Figure 2-23). These instruments control the pressure in the KO Drum. PIT-100 senses the pressure in the line and transmits a pneumatic signal to PIC-100, a field mounted pneumatic controller. PIC-100 develops the correcting signal and transmits it to PV-100, a 6" butterfly valve. Upon loss of its air supply, PV-100 will fail to the open position.

Third, as shown in Figure 2-24, Level Loop L-100, there is an electronic level loop consisting of LT-100, LIC-100, LI-100, LY-100 and LV-100.

The LT is shown as a displacement type level transmitter, LT-100, which sends an electronic signal proportional to level to the distributed control system. The DCS includes a software configured indicating controller, LIC-100. The level signal transmitted to the DCS also runs through a local electronic level indicator, LI-100, shown by a note to be located at the bypass of control valve LV-100. It might be necessary to use LI-100 if control valve LV-100 is out of service for maintenance while the electronics of the DCS are still in service. A process operator could manually adjust the drum level by positioning the bypass valve while watching the drum level readout on LI-100. LY-100 is the I/P converter. LV-100 is a butterfly control valve that fails closed upon loss of its air signal, since the power to the valve actuator is only the instrument signal. The level indicator, LI-100, is powered by the signal wires, or loop power, since there is no separate power source shown on the P&ID connected to the indicator.

Figure 2-24: Level Loop L-100

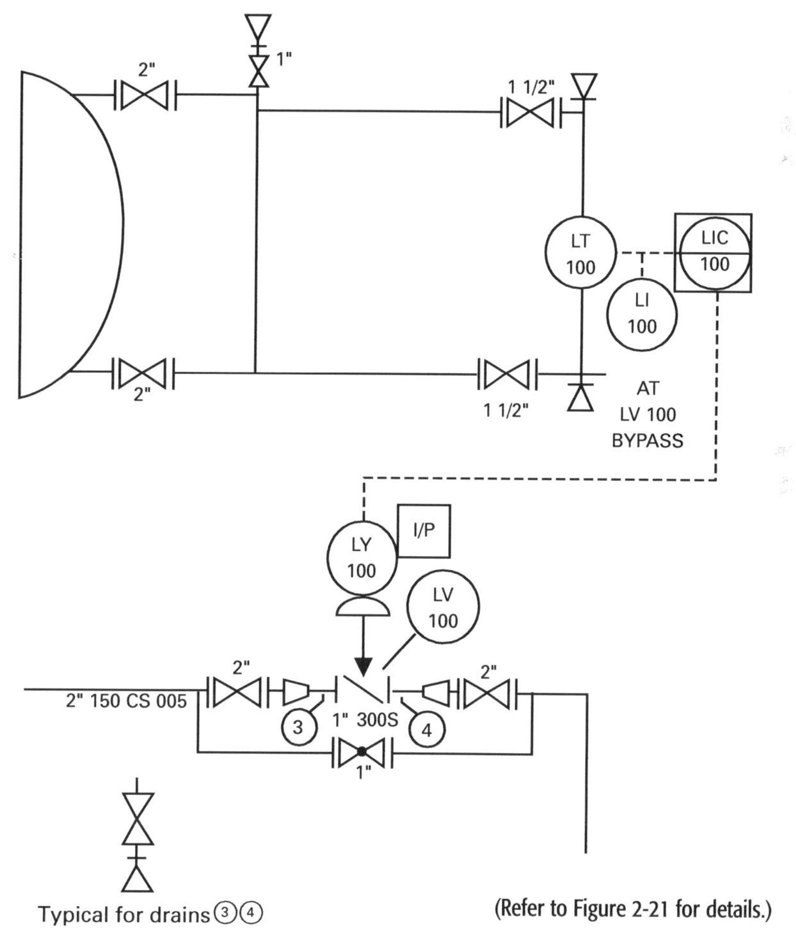

(Refer to Figure 2-21 for details.)

The piping design includes 1" plugged drain valves upstream and downstream of all control valves to drain any residual liquid in the line when valve maintenance is necessary. In use, the drains are connected to a portable recovery system. After the liquid is drained, the control valve may be removed without atmospheric contamination.

The level devices are all connected to a strongback, also called a pipe stand or instrument bridle, connected to KO Drum 01-D-001. A strongback is a continuous pipe, typically 2" nominal, connected to a vessel through the top and bottom block valves. It is shown as the vertical line just to the right of the valves. The level instruments are mounted to the strongback and include the level gauge (LG-1), high level switch (LSH-300), low level switch (LSL-301), and level transmitter (LT-100).

Level devices are often, but not always, connected to vessels in this manner for several reasons. First, attaching level instruments to an isolatable strongback, or instrument bridle, facilitates testing the instruments without disturbing production, or without opening the entire vessel. If suitable for the process fluid, the strongback can be isolated, the vent valve can be opened and a process-compatible fluid can be introduced through the drain to check the function of the level devices.

Second, vessels usually have long delivery times and therefore are purchased early in the project. The level instruments are normally purchased much later. Accurate dimensional information for installing the level instruments may not be available when necessary to be able to install the connections directly on the vessel. Therefore, in this example, two 2" connections are placed on the vessel for the strongback. The instrument connections to the strongback may be scheduled much later in the project. Third, vessel connections are more expensive than piping connections, the two strongback connections are fewer than the numerous connections needed for the gauge, transmitter and switches.

Figure 2-25: Local Panel Switches & Lights

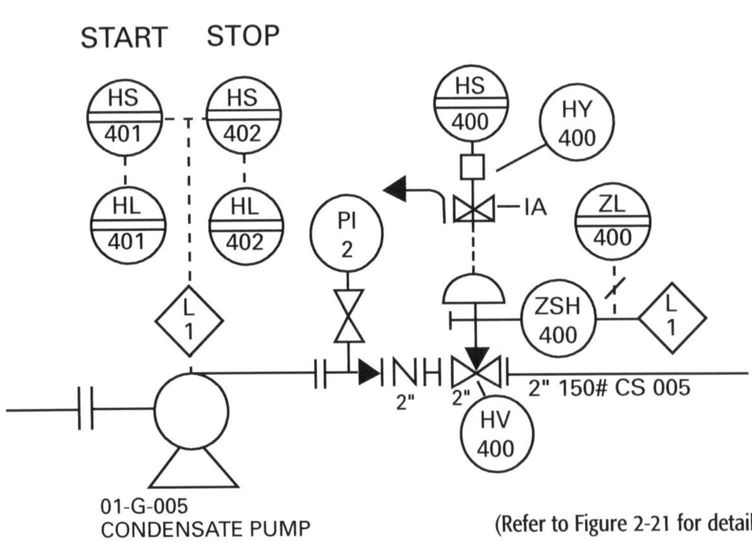

(Refer to Figure 2-21 for details.)

The authors do not advocate connecting all level devices to strongbacks rather than directly to the vessel, but we are showing this type of installation as a possibility.

In Figure 2-25, the buttons (hand switches) and lights (HS/HL 401 and 402) are used to start, stop, and indicate the state of pump 01-G-005. The tag marks are located in a circle with two parallel lines. This tells us these hand switches and lights are located on a local panel or, as stated in ISA-5.1, "Located in or on front of secondary or local panel or console." This local panel might also be the motor control center. An identifier could be added to indicate which panel contains the components. Usually a three-letter acronym is sufficient to call out the panels. These acronyms must be identified on the legend sheet.

HS-400, also on the local panel, energizes solenoid valve HY-400. This opens HV-400, a pneumatically operated on-off valve. The tee symbol (T) on the actuator shows that there is a manual operator on the on-off valve. A manual operator permits operations personnel to override the pneumatic signal and close the valve.

As shown by the diamonds with the internal L1 legend, there is some on-off or binary logic control, relating to 01-D-001. The P&ID shows that the on-off control involves hand switches (HS 401/402), position switch high (ZSH-401), and level switch low (LSL-301). See Chapter 6 for more information.

Returning to Figure 2-21: The safety valve (PSV-600) at the top of 01-D-001 has been sized, and a Specification Form prepared by a qualified designer. The design, size and location of safety valves are sometimes the responsibility of the control system group, sometimes of the process design team, or sometimes of the mechanical designers. The responsible group must ensure that safety valves are shown correctly on the P&ID.

There are two pressure gauges shown on the P&ID: PG-1 and PG-2—one to indicate the pressure in the vessel and the other to indicate the discharge pressure of pump 01-G-005.

To minimize contamination, the strongback drains into an OWS, an oily water sewer. This is a separate underground piping system that is connected to an oil recovery system. A symbol for an oily water sewer is shown as the "Y" with the OWS label.

A "typical for drains" sketch is shown on Figure 2-21 for the seven drain valves, upstream and downstream of control valves FV-100, LV-100 and PV-100, plus one to drain the strongback.

Summary

In this chapter we looked at P&IDs. The letters stand for different names. P refers to piping or process, the I for instrument or instrumentation, the D is for drawing or diagram. P&IDs are sometimes called by other names, for example, engineering or controls flow diagrams or sheets.

P&IDs describe a process using symbols and numbers to specify an instrument tag number. We use the symbology in ISA-5.1 as a definitive reference. We explained the meaning of some of the symbols, but not all. P&IDs also show equipment, piping and electrical devices. The chapter includes a discussion of how the P&IDs develop and their role in the design of a process project.

In general we have looked at P&IDs in depth describing what information might be shown on a P&ID and what form that information might take.

CHAPTER THREE

Lists, Indexes, and Databases

Overview

Commencing with the advent of computational technology in the 1960s, lists and indexes documenting the mechanical equipment and control system devices that make up process facilities have evolved in increasingly useful ways. The pre-digital typed or handwritten lists were first developed to organize specific data for the construction and operation of complex process facilities. These documents included, as a minimum, equipment and instrument lists. There were other lists as well: piping, motor, cable and circuit schedules and so forth. These all served the most basic functions such as defining component identifiers so things are identified the same way on all the drawings and in the field, such as the ISA tag number or with an equipment number. They also called out critical component requirements such as an instrument range, or the horsepower, energy, throughput or volume, all in a tabular format. The "list" became an "index" when digital sorting allowed use of reference information such as the physical device location and maybe the specification call out. The utility of lists and indexes is in organizing information and presenting it in a way that is easy to access.

There is no standard for the format or content of lists and indexes; they are set up to suit the user's requirements, local practice and maybe tradition. The term list and index tend to be used interchangeably; however, an index is properly a list augmented by references to other associated data such as a location drawing, an installation detail, a purchase order, the specification, or other relevant device specific data. When they were hand printed, their usefulness was somewhat limited because they were difficult to sort. With even basic computer technology, the ability to sort and focus on certain data within the list increased their usefulness and made them more critical to the construction and operation of the facility.

The equipment list usually includes all the pumps, compressors, vessels, tanks, air handlers, and so forth. The control system devices are listed on an "instrument list". For each, data in addition to the equipment or instrument identifier and title is added to suit requirements for that facility or project. There is also likely a loop list that captures the name and number of all the assigned loop numbers to prevent duplication.

Today, in a digitally integrated process plant, data developed for a specific purpose such as an instrument list is then available for shared use by other software applications, including those for control systems. Each list or index is then just a formatted presentation of a subset of the available digital information that defines critical information.

Increasingly, all the data is collected in a master digital file called a database. The information in the database is then available for use for the drawings and specification as mentioned, but also for other tasks unrelated to the design and construction of the facility, such as maintenance and the asset management system. An asset management system is a software platform used by maintenance staff to track the health of components and to more efficiently plan the maintenance and replacement of the "assets" of a facility, the equipment and even control system devices.

Instrument Lists were first used to compile data, usually for a limited number of specific tasks, such as keeping track of the tag numbers assigned or the individual components that had to be purchased. The data was manually copied from the source documents such as P&IDs and specification data sheets to the list using a (now-extinct) manual typewriter or by a control system drafting technician with legible writing skills. Legible penmanship actually existed back in the dawn of time, in the '80s.

In the maintenance shop, calibration cards were scratched out using some of the same information, using source documents such as the P&IDs, specification data sheets and copies of purchase orders. Possibly, over at the DCS configuration desk, I/O (input/output) was assigned using much of the same source documents onto a smudged, handwritten termination list which was then transferred again to a preliminary Loop Sheet. This manual location and transfer of information took time and introduced errors, but it was the most practical method, given the available tools.

Indeed, some facilities continue to maintain instrument data the same way, albeit with some assistance from technology. Perhaps they have gravitated from the old manual typewriter to a computer-based spreadsheet program or even a database. Electronic spreadsheets and databases are hugely useful, given the ease of revision and the ability to sort data to your heart's content. On projects involving multiple entities such as a design consultant, contractors, a system configurator and the plant maintenance staff, the ability to email the files from one party to the next for revision is very helpful. Some programming and configuration software allows the importation of data from other commonly used files such as spreadsheets and databases. This practice alone eliminates common transcription errors and provides overall increases in speed, efficiency and accuracy. The ability to easily sort on similar devices or tag numbers simplifies error checking as well. This system works well for many facilities, but in this chapter you will discover more opportunities for improvements in data handling and use.

Figure 3-1: Instrument Index is a brief description of a "traditional" Instrument Index or List.

Figure 3-1: Instrument Index

ISA-5.1 describes an Instrument Index as follows:

An Instrument Index should contain references to all instrumentation data required by owner and/or government regulatory agency management of change requirements and contain, as a minimum for each loop:

a) Loop Identification Number.
b) Service Description.
c) Instrument Identification/Tag Numbers.
d) Piping and Instrumentation Diagram drawing numbers.
e) Instrument Data Sheet numbers.
f) Location Plan drawing numbers.
g) Installation Detail drawing numbers.

From ANSI/ISA-5.1-2009

Instrument Databases Boost Efficiency

Today, information developed and maintained for process control tasks will assuredly exist in digital form, all of which can be collected in what is essentially an infinitely flexible data set – the instrument database. The traditional Instrument Index is simply a subset of the information available in the database.

Today's instrument databases have the capability to be the primary repository and source for data rather than a copy of data generated elsewhere, which was the old method. Data can flow "out" from a single entry in a database to all other documents, rather than "in" from the documents into the database. Software programs can link directly to the database so there are no transcription errors. Some software will also identify changes to notify the user when there is a difference between the source data, the database, the application, and possibly the DCS configuration program. For example, if a suffix was added to a control valve tag after a new control valve was added to a loop, software can immediately, efficiently and accurately compare the source data to the application. The goal is always to write (and check!) the primary source data once, then use it many times.

History

Instrument databases contain information that is referenced by different software packages to produce maintenance schedules, calibration records, Instrument Indexes, Loop Diagrams, text on a P&ID, and configuration files for the process control computer.

Prior to the explosion of computer use, everyone manually typed or wrote lists of the many devices to be purchased and installed in some simple, chronological order. Field instruments were mechanical or pneumatic in nature, with little electronic computer capability.

When mainframe computers first became accessible, users key punched data cards to describe the instruments and, after some programming effort and some time delay for shipping off the cards and for the return of "computer runs" (printouts) from the computer center, users were able to sort the data into presentations that were more useful, they developed specialized lists to simplify design, maintenance and other activities.

Engineering and construction contractors developed some of the earliest applications of computer-based Instrument Indexes. Their work required tracking the progress of the control system portion of the project; the specification, bid, purchase and delivery of instruments; the control system devices. Programs were written to allow construction progress reporting. Sharing the resulting lists beyond the company that prepared them was difficult since different engineering, construction or operating companies used proprietary programs and dissimilar software. Also during this embryonic computer era, instruments were becoming increasingly self-capable, stand-alone devices that could not yet communicate well with other devices. That too will evolve.

With the advent of desktop computers, even non-programmers were able to prepare and manipulate data with relative ease. Simultaneously, the development of computer chips also allowed field devices to become increasingly "smart," "smarter" or "intelligent" (depending on how they were described by the company selling them). Along with intelligence now residing in a field device, options appeared in how the device was configured, so the amount of information that had to be developed and archived to define the device increased. The additional features not only increased the flexibility of the device, they also allowed previously unheard-of accuracy and stability in the signals transmitted. Luckily, the desktop computer and database or spreadsheet software were available to manage the additional information. As field devices became more complex and capable, the tools to manage them became more capable as well.

For Instrument Indexes, in particular, the single most important technology development may well be the relational database. The *ISA Dictionary* defines a relational database as "An information base that can draw data from another information base situated outside it." In practice, a relational database is a set of tables that allows data to be used in an infinite variety of ways, without re-entering the data. Furthermore, the ability to share data among different software packages continues to improve, so information in the database can be used by the myriad of software packages used in a process facility.

It is important to remember this feature of an instrument database: To make the best use of the powerful capabilities of the software, information should only be entered once. Once it is entered, it can be used in many different presentations, indexes or reports designed to suit the requirements of the user. For example, the maintenance technician may want calibration reports or component replacement information, the person doing the control system configuration may want function identifiers and I/O assignment data, and the purchasing department may want specification information and delivery status. Construction staff needs to know how many instruments were installed and how many have yet to be installed. The capabilities are awesome.

Control Systems Group Manages Device Documents

The control systems group handles information about many individual devices. It is arguable that this group handles more unique pieces of information and more individual devices than any other discipline in a process facility or industrial design company, with the exception of business departments. The discrete bits of information are not particularly complex, but the devices are numerous and information about them appears in many different documents with many different uses.

Consider for a moment a simple instrument tag number—our ubiquitous pressure transmitter PT-100. The device appears first on a P&ID, then later on a Loop Diagram. Its tag number also appears on Specification Forms and in purchasing documents: a request for quotation and a purchase order. The plant instrument shop maintains a device log for PT-100—listing, for instance, the calibration range, the manufacturer and the model number. The tag number appears on the Location Plan. There may be an I/O Loading document that defines the termination point for the instrument signal. The construction status of the instrument was probably recorded and updated daily during initial installation of the control system. An Installation Detail was assigned to or developed for each instrument to show the installation requirement of the device.

Suffice it to say there is a lot of data associated with that single component. There may be several lists and indexes to record and present this data. A modern facility can assemble all this information in one place: an instrument database. Figure 3-2 shows the complex array of information that is managed by one company's instrument database program.

As mentioned earlier, in its most basic form instrument data was originally presented as a simple list: "Here are the instruments I care about," with little additional information or sorting. The order was sometimes chronological, so its usefulness may have been limited to ensuring that all the devices had been specified, purchased and installed. Finding a specific device meant you had to scan the list for it or know when it was listed. This minimal list approach is still used today, often with rudimentary sorting, and most often as a checklist. It offers little information to interest the maintenance group or to use as a design tool.

With the addition of a bit more information, the function of the simple list may be expanded to meet other needs; to add external information. Fields containing the Specification Form number, purchase information, and installation and commissioning milestones provide effective support for the construction effort by summarizing the status of the instrument installation portion of a project.

The most common form of instrument data list—the Instrument Index—arose with the addition of drawing references. The Instrument Index provided a cross reference from the tag number to the drawing numbers for the P&ID, the Loop Diagram, the Location Drawing, and all other relevant documents. Today, the terms are used interchangeably: instrument list, instrument index and, in today's computer-based work, the instrument database.

In its most complex form the instrument database may contain information that is limited only by the imagination (and the budget) of the people that use the information.

To List or Not to List: That is the Question

To be useful, an instrument database should include all the tag-marked physical devices that comprise the measurement and control system. The general rule is: if something has to be purchased and then mounted, wired, tubed or calibrated, then it should appear in the instrument database. The corollary to this rule, by the way, is that these devices should also appear on the P&IDs, since the P&IDs are the source documents for the database. The untagged control room operator stations, under these criteria, would then not appear in the instrument database. However, some, or maybe even most, design or construction companies will make up a tag number, specification sheet and database listing for these as well, since control system device tracking is otherwise completely integrated with the instrument database. If listing general components like operator stations suits your purpose, go ahead and list them. Remember to always be consistent in listing equipment. If you tag and list one operator station, you should tag and list them all.

Deciding what information to include in an Instrument List/Index/database can be contentious at times. For example, spirited discussions can take place within process control design teams as they decide if software functions are to be included in the instrument database and Instrument List. It is useful at such times to establish criteria to help current and potential users decide which information will be maintained within the database. Ask these questions:

1. Is there a specific need for the information, one that is not better met by something else? Don't list data just because it is available. It has to be needed by someone that actually uses the document.

2. Is there value added in having this information in the database?

Listing each data record has an associated cost. The software may have to be modified to carry the information. The data has to be found, entered into the database, checked and printed. Once entered, data in the database has to be maintained, so be sure to consider the cost of maintaining it. Having out-of-date or inaccurate data is WORSE than having no data. If there is no entry in the database, then someone will go look elsewhere for the information and will probably find it to be both current and accurate. The person or entity entering the data does not get the final say in the value, the "customer" or user of the data should make the decision.

Ultimately, a designer's customer is really the maintenance staff and the operators. Their needs must be integrated into the standards.

Enter data ONCE and only once. If the same information is entered twice, entry errors will seemingly always cause the information to be wrong in both places!

An Instrument Database Success Story

Figure 3-2 shows one company's instrument database system. It is a complex one that was developed over several years to meet ongoing challenges – and it has proven to be very successful in efficiently maintaining information accurately. One of the primary developers of the system, Ken Brabham, PE, offers some insights into its development and application.

"The necessity for our database architecture arose for the following reasons: to reduce the drawing count, minimize data entry time, to improve accuracy and to complete projects in less time. This approach has now been used on many projects. A typical control system for this database consists of distributed PLCs with a centralized SCADA system. Project size is normally in the range of 3,000 to 5,000 real world I/O points.

"The database was originally developed to handle design information – primarily to generate construction documents like data sheets and wire schedules. The schedules were written and presented in a format designed to replace manually drafted wiring diagrams, thereby eliminating hundreds of I/O drawings for each project. The database is linked directly to the P&IDs so when a drafter adds or modifies an instrument tag on a CADD P&ID, the system includes software to automatically reflect the change in the database. The system became more than just a database of information – (it became) a design tool for all projects.

"This database not only became an invaluable tool for the design phase of our work, but as projects transitioned from design to the systems integration phase, we found many additional uses as well. One significant use stems from my inability to type at a measurable rate! My slow keypunch speed led me to hone my database programming skills, so I now hardly need to type at all. Since the PLC and SCADA software packages support data importing and exporting, all that's needed is to simply reformat the design data so it can be directly imported into the various control systems. This way, the PLC programmer never has to enter any of the real world I/O information. He or she simply imports the information as needed: the instrument tag number, point description, I/O assignment and ranges for analogs. This greatly reduces the amount of time spent digging through drawings and specifications to obtain such information.

"When the PLC programmer completes programming on a system, the data is exported and once again reformatted for importation into the SCADA package. However, this time all SCADA points can be imported, not just the real world I/O points.

"Now both the PLC programs and the SCADA points have been generated from the design database without the need to retype tags, descriptions and addresses. In fact, the instrument tags had only been typed once by the drafter during the P&ID development phase, and that information was automatically extracted and imported into the database. Since all of this was obtained directly from the design database, human errors are greatly reduced. We also discovered that once you maintain this information in a database, you continue finding additional benefits like error reporting for duplicate tags or duplicate I/O assignments. We have even used the database system to track individual instrument status from procurement through installation and acceptance.

"What's the downside to a database approach, you ask? Well, for me, I still can't type." – Ken Brabham, PE, Unpublished Notes

Figure 3-2: Typical Example of a Company's I&C Data Flow Diagram

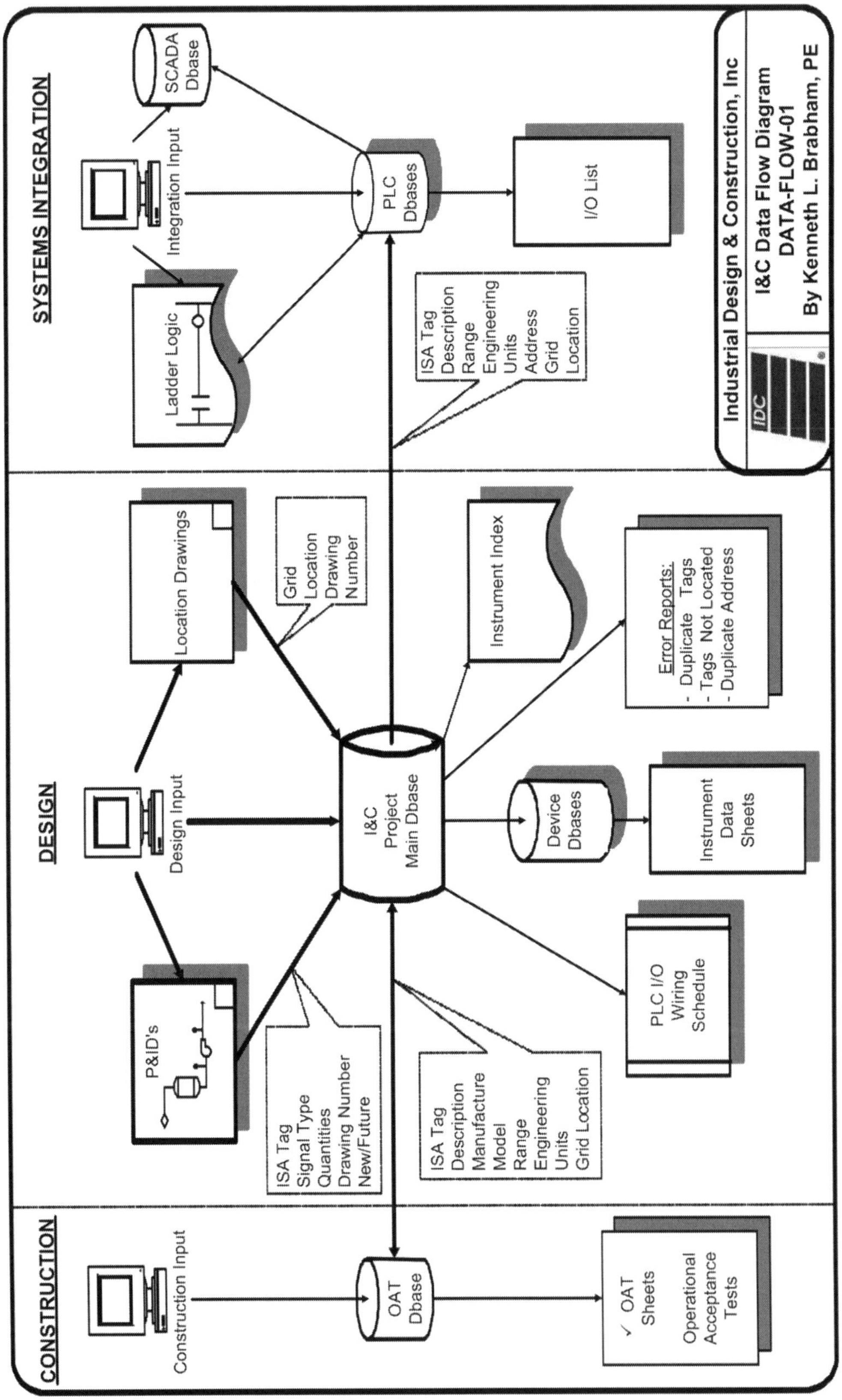

The power of the data and database reports and presentations should be used! Encourage widespread use of reports and data linking. If no one uses the database, there was no value in making and maintaining it. In the authors' experience, increasing the number of people using the database will increase the quality of the data, as long as the people using it are careful. It is also not particularly difficult to mess up the information in a database, so revision control and backing up the data daily are important.

You should develop a system to verify the data upon entry, to back up the data electronically, and a method to track changes to it, and you should regularly distribute the resulting reports. The key is to have a system, in other words an agreement on procedures, and you should have someone who is assigned to be in responsible charge of the database. Anarchy and databases don't mix.

Any discussion regarding the need to list software functions (Control, Ratio Control, Feed Forward, Sum, Integrator, Record, Trend, Bias, etc.) in the database should probably consider whether or not a master list would be of use to the people preparing the control computer configuration. For instance, the software functions might be an important review and approval item between the designers and the operators and maintenance staff, so there could be advantages to including the functions in the database. Listing them could simplify the approval and progress report development. This is an example for discussion because listing software functions in the instrument database is not a common practice.

Figure 3-3: Instrument Data Fields

Basic Data
- Tag Number
- Function and Service
- I/O Type (AI, AO, DI, DO)

Technical Data
- Calibration Range and Units
- Rating
- Power: Loop or 120V AC

Index Data
- P&ID
- Loop Sheet Number
- Installation Plan Number

Connection Data
- Junction Box
- Marshalling Panel
- I/O Rack
- Address

Maintenance and Operations
- Stores or Stocking Number
- Calibration Date
- Calibration By
- Manufacturer
- Model Number
- Vendor
- Vendor Phone Number
- Purchase Order
- Original Delivery Date
- Current Delivery Date
- Received – Yes/No
- Receiving Report Number

Construction Data
- Calibration or Shop Approved
- Issued to Contractor – Who?
- Issue Date
- Installed Date
- Tubing Complete - Date
- Wiring Complete - Date
- Checkout Date
- Commissioning Date
- Turnover Date

Commonly used fields of information in an instrument database are listed in Figure 3-3.

The data fields listed in Figure 3-3 are only suggestions. Fields that you include should be selected to meet your facility's requirements. As mentioned in the criteria above, there should be some value returned to users for maintaining the information. If no one at your facility needs to know whether a control valve has an open or closed yoke, then there is little need to maintain that information in the database. The more useful the database is to users, the better the information it will contain. The more esoteric and complex the information contained therein, the less easy the database will be to update and maintain.

Construction information may appear in an instrument database, as shown in Figure 3-3. In fact, procurement or construction activities are likely to initiate its development. The maintenance database used five years from now probably will have started as a tool to monitor procurement, deliveries and construction progress. Yet, after construction is complete, that information will probably never be referred to again. In a relational database, construction information can be maintained in a separate table that can be abandoned after the facility is turned over to maintenance and operations. At the same time, a maintenance-oriented table of information will come into use. Calibration and maintenance work will be scheduled and performed, and any changes to the instruments will be recorded in a database for the remainder of the facility's life.

Database Becomes Master Document

The instrument database, or Instrument Index, is, or should be, the master document used to record and maintain tag numbers and function descriptions. This data may appear in many places including Specification Forms, Loop Diagrams and possibly on the DCS screen.

The function description is sometimes referred to as the loop title. It is much less confusing, not to mention professional, to have all the text fields read the same way wherever they appear. The instrument database ensures that this happens. The function description field should be written only once, no matter how many times and no matter in how many presentations it is used. A relational database allows the data to be pulled from the same field every time it is used.

The loop service text field describes in words the function of the loop. Having one master document for this information is critical to prevent duplication of tag numbers and to maintain a pattern for your service descriptions. More than one instrument has been tagged and purchased because the title appeared twice using different descriptions. A fixed pattern will make the text easier to understand and more efficient to develop. Service descriptions are important from a practical standpoint when someone is trying to locate a specific loop while using the Instrument Index or when thumbing through a pile of Loop Diagrams.

There are essentially two basic formats used for service identifiers, with subtle variations on those formats. The first is to put the loop function first, followed by a "for whom" or "for what" description. For instance: "Level - Tank A" or "Flow - Chilled Water to Condenser."

The hyphen is just a delimiter for clarity, but the delimiter and its surrounding spaces should be established at the onset of the database development, since the delimiters and spaces can impact electronic sorting. Also, aesthetics are important, using the same delimiter and spacing will look tidier.

The second way to call out the service is to conversationally state what the loop does, for example: "Tank A Level" or "Condenser Chilled Water Flow." The conversational service identifier is more difficult to sort by function, but it is probably more natural in use. It is important that you don't mix formats; once you pick your format, do not allow deviations.

Figure 3-4: Common Service Description Formats

Function Led	Level – Tank A
Conversational	Tank A Level
Function Led	Flow – Chilled Water to Condenser
Conversational	Condenser Chilled Water Flow

Ensure Agreement

This may seem like a minor issue, but it is important to work with the operations people to ensure that the equipment titles you use are the ones with which they are familiar. This is especially true with equipment upgrades or replacements. Calling a tank the Secondary Hot Water Tank may be perfectly correct, unless everyone for the past 35 years has called it the "Green Tank." When the maintenance control system technician is called in at 3 a.m. to fix the level loop on the "Green Tank," he or she won't really care that your equipment title is now more technically correct. However, they will care that they had to search around for 20 minutes to find the troubleshooting documents for a tank that you helpfully re-identified after 35 years of what you perceived as an error. They will share their experiences with you to encourage your professional development; in other words, you're gonna get yelled at. The authors know this because, well, one of us did it.

When you are part of a design project, it is imperative that the equipment titles and equipment numbers used in the instrument database are the same as those used by the equipment and electrical groups. This is the power of the relational database. The instrument database can link, or relate, to the master equipment list. Titles used in the instrument database must agree with those used on the rest of the project. It is hard to find a ventilator to add a proof-of-flow switch when everyone else called it a fan.

When developing tag numbers and function descriptions, it is good practice to check the format requirements of ***all*** the control and monitoring software programs that use the information. Establish the character counts and formats used to make up the text strings in all the documents for compatibility. The Instrument Index (instrument database) can then provide the template for creating the text that ensures the correct use in all the applications. For instance, your DCS may limit character fields for a service description associated with a tag number to 32 characters, so your service text must follow that format. This is easy to set up in an instrument database.

Linking Information

The Instrument Index is a core document prepared by the control system group. A good Instrument Index is a summary of the physical pieces that make up the control system. The Index has undergone something of a renaissance with the expansion of computer use and the inter-operability of software. There are many different applications that can make use of information stored in a software package. With the computer skills contained within today's control system groups, it is only natural that some very useful information-sharing has arisen offering increased efficiency, speed and accuracy for designing and maintaining control systems.

For example, some firms have linked the computer aided drafting software with the instrument database so the P&IDs and the Location Plans directly and automatically feed their information for use by the full suite of instrumentation and controls documents such as the instrument index, specification data sheets and even control software. ANSI/ISA-5.06.01-2007 provides examples of control software basic, I/O, interlock and operation data that all starts with the information drawn from an instrument database which is, in turn, populated using data on P&IDs as well as other drawings and specifications. The capability now exists to not only extract the instrument tag number from a CAD-based P&ID, but also the associated piping line number, P&ID number, equipment number and even the I/O type with clever use of rules programmed into the software to look at the tag number and even the type of signal line that is drawn on the P&ID at the instrument symbol. Indeed, with many CAD systems, any information that appears on a drawing is available for extraction to other applications, assuming compatibility can be found between the different software systems.

Since Location Plans are drawn to scale, it is possible for software to find a tag number on the drawing and feed back to the instrument database a grid reference location for that device. This both simplifies development of the instrument database and increases its accuracy. The tag and equipment numbers will be exactly as shown on the drawing, since they use the same information source.

Not only can a software tool automatically and accurately link the tag number and a physical location, but it provides a way to verify design progress; if the software is unable to find a control system component's location, then the designer has some more work to do as the device is not yet located.

This software link between the database and the drawing is not limited to the Index, Location Plans and P&IDs. This same process can be used to confirm that each measurement and control component has the appropriate input or output (I/O) assignment within the control system, and that that I/O point is shown on the appropriate drawing or table. The drawings showing these I/O assignments are referred to by many names such as card loading diagrams, card sheets or I/O lists. Essentially these documents all serve the same purpose – to show where signals associated with measurement and control devices are connected to the control system and, importantly for ongoing operations, the I/O points that remain unused and available for future use. As with the Location Drawing, if no I/O assignment is returned from the drawing or table that define the connection points, then that device has not yet been assigned.

There are challenges and pitfalls, of course. The linkages between software applications and the manipulation of the data probably will require custom software which can be expensive to develop and maintain. Your time savings and increased accuracy must offset the cost to develop the tools. The "information mining" software might require that drawing data be entered in a particular way. It may be necessary to train staff to manipulate the computer aided design and other software to enable entering and extracting information in a useful manner.

Software Development Challenge

The biggest challenge with any instrument database is developing or acquiring the software. Third-party software is usually available to perform the desired tasks but it comes at some expense, both initially and in yearly licensing. Third- party software allows quick startup of your instrument database without having to learn the language of databases. Third-party software may also offer the advantage of continuous technical support independent of your own staffing for as long as the facility can (or will) pay for the support.

You can write your own software in-house. Many companies have spent significant money developing software to perform a specific task, only to have the programmer depart, leaving the software package unsupported. This loss of support can happen with third-party software also, so care should be taken to pick a reputable, respected company that will be around five years from now. If you are responsible for selecting the software, you should get references and call the supplier. It may also be useful to look at the supplier's software development history to get a sense of how they handle upgrades; compatibility is good.

Happily, there is a trend in computer use that bodes well for using databases for Instrument Indexes and reports: we are all becoming much more comfortable with some pretty complex computer software. As more high schools teach software classes, skills to manipulate data in meaningful ways are becoming more universal. A software package that looks impossibly mysterious and daunting to someone who left school before 1980 may look like, well, just another type of a pencil to someone in school right now. If you are challenged by the daunting task of learning to use the software, turn it over to the kid in your group; she or he was messing with databases in high school. As the software becomes more familiar, manipulating data in a relational database is becoming simpler as we all become more experienced with the available tools.

CHAPTER FOUR

Specification Forms

Introduction

When the control systems group adds device symbols and tag numbers to a P&ID, they simultaneously undertake two other activities. First, they enter the tag numbers into an instrument index, list, or into a database, using this to ensure that the tag numbers are unique. Second, they prepare written definitions of the instruments so the parameters under which the devices will operate are known for the programming and configuration of the SCADA, distributive control systems, programmable logic controllers, and personal computers. With this information the device can be purchased. This information is most often, but not always, on a Specification Form, sometimes referred to as a data sheet or spec sheet. SCADA is defined in the *ISA Dictionary* as supervisory control and data acquisition – a generic name for a computerized system that is capable of gathering and processing data and applying operational controls over long distances.

Clearly, the information available on a P&ID is insufficient to completely define the instruments, so additional information and a way to present it are needed. Referring back in Chapter 2 to Figure 2-21, the symbol and tag number define FT-100 as an electronic vortex shedding flowmeter. However, this is only part of the required definition. The programming and configuration designers don't know the signal or the range, they may not know how the device is powered, they won't know whether the signal is sourced in the field or it is loop powered nor could a supplier accurately quote a price and delivery for the device. The supplier needs to know, as a minimum, the process fluid to be measured, the pressure and temperature rating of the system, the allowable piping connection information, the power and signal requirements of the device, the electrical area classification, and the housing rating. To take full advantage of the supplier's expertise, the supplier also needs to be told the minimum, normal and design flow rate, the operating pressure and temperature of the process, additional process fluid information, including viscosity, and the device's metallurgical and materials requirements.

Since a great deal of information is required to define each device, a Specification Form is used. This aids in organizing the information and ensuring that complete and consistent information is acquired. Blanks on a Specification Form imply missing information. The need for uniform Specification Forms is stated in ISA-20-1981 *Specification Forms for Process Measurement and Control Instruments, Primary Elements, and Control Valves,* as follows:

"Because of the complexity of present day instruments and controls, it is desirable to have some type of specification form to list pertinent details for use by all interested parties. General use of these forms by users and manufacturers offers many advantages as listed below:

1. Assists in preparation of complete specification by listing and providing space for all principal descriptive options.
2. Promotes uniform terminology.
3. Facilitates quoting, purchasing, receiving, accounting, and ordering procedures by uniform display of information.
4. Provides a useful permanent record and means for checking the installation.
5. Improves efficiency from the initial concept to the final installation."

The preceding quote was written specifically for ISA-20, and it makes the case for the use of uniform sets of Specification Forms. Figure 4-1 provides a definition of a Specification Form.

Figure 4-1: Specification Forms

- **Specification forms**, or **data sheets**, define the tag-marked devices that make up a control system
- The form contains information necessary to secure vendor quotes and to purchase devices

The development and management of data needed to complete Specification Forms can be a significant part (perhaps 25%) of the control system designer's work-hour budget. The specification of instruments can, or should, be a point of significant discussion between the designers and the maintenance team. Specification Forms support that discussion by listing the important features of an instrument.

There are many variations of Specification Forms. Most engineering contractors have developed a set for their use and some instrument suppliers have their set. Many of these sets are based on the ISA-20 standard. ISA also has a newer technical report, ISA-TR20.00.01-2006 *Specification Forms for Process Management and Control Instruments–Part 1: General Considerations* (See Appendix C). It is also available in Microsoft Word Format on CD-ROM.

Of course, everyone prefers to use the forms with which they are most familiar. An ISA Specification Form modified by a third party typically will present the data in a similar format, but will likely drop fields not applicable to the typical devices used in a particular industry. For example, the ISA form for pressure transmitters includes sections on controllers and chart recorders that are not applicable to intelligent (smart) electronic transmitters, so the form used by some companies will drop those sections and add fields needed for the devices being described. A modified Specification Form may include a definition of the communication bus used, plus a place to list the information needed by the device supplier to pre-program intelligent transmitters before shipment. Some owners opt to list calibration or range verification and acceptance signatures on the same sheet, so the specification form includes the installation record for posterity.

The intent of all of the form's variations is the same. A Specification Form defines the instrument completely and accurately so vendors may quote on, and supply the device so construction personnel can install and check it out and so maintenance personnel can properly perform calibration, range verification and repairs.

ISA-20 consists of Specification Forms for 26 commonly used instruments and—something that is often overlooked—provides a line-by-line explanation of the information intended to be furnished in the data fields. This is an invaluable reference, particularly for those who may be less experienced in working with Specification Forms—and they are a pretty good review for those who have been working a while in other areas. ISA-TR20.00.01 is a newer set of Specification Forms issued by ISA as a Technical Report. The new technical report contains forms that have been modified and expanded from the original ISA-20. There are entirely new forms for instruments not included in ISA-20.

Specific Knowledge Required

When filling out a Specification Form, the control system specialist must have at least three types of knowledge:

1. **Process Knowledge:** The specialist must know how the process works and how it can be controlled. The specialist must know the process flows, process pressures, process and ambient temperature and, very importantly, any expected variation in the process conditions. This person must know the effect process fluids will have on all instrument wetted parts and they must understand the effects of exposure to corrosive or hazardous materials or extremes of temperature. Corrosive materials might cause the device to become inaccurate, inoperative or dangerous; if such an instrument is in an hazardous area it might cause an explosion or fire.

2. **Codes, Specification and Standard Knowledge:** The specialist must be familiar with requirements set forth in local and national codes. In the United States, these include National Fire Protection Association (NFPA) Standard 70—the National Electric Code (NEC). The specialist also needs to know the specialized codes of the industry (see Chapter 10) in which they work, and, with increased globalization, they need to be aware of offshore standards that the owner might add that might be more stringent than codes of the industries involved. (See Chapter 10)

   The specialist must know the specifications and standards that are to be followed. Many operating entities and engineering contractors have comprehensive specifications that dictate the quality of the materials, equipment and documentation to be supplied for a project. There are also many industry standards that may be relevant and must be followed.

3. **Product Knowledge:** The specialist must also know a great deal about instrument suppliers, including which vendors can supply each type of device. When competitive bids are issued for instruments, the specialist needs to know what features are common to all devices, regardless of vendor. It is also necessary for the specialist to recognize that features presented as "hot selling points" might be made to seem important by a particular salesperson, but may not actually be required or welcomed by the people using the instrument.

Looking at a simple example of this issue: XYZ Company has a new transmitter with a guaranteed accuracy that is ten times better than the competition's. This is interesting to know. Understand, though, that if higher accuracy is made a requirement by its being listed on the Specification Form, that will prevent all but this vendor from supplying the device, preventing competitive bidding. Is that accuracy actually needed for the application? Will it provide some significant benefit in the control of the process? One of the tasks of the specialist is to know which features are really required. For a competitively bid project, a specification that needlessly limits the number of prospective suppliers will result in higher costs for the devices. The other side of the argument is that a specification that is too loose will result in poor operation of the plant.

Alternative Means of Definition

Up to this point we have concentrated on the Specification Form as the means of defining instruments. Other means can be efficiently used to specify the requirements for procurement and installation, in particular when the components are "off the shelf" or "bulk" material items. In other words, these items are "one size fits many" components that do not require specialist knowledge for selection and installation. Examples of these components include local gauges, gauge isolation pigtails, condensate traps, and even pulsation dampers if they are not integral to an associated instrument.

Another advantage of listing these components generally or as bulk items is that if you don't have to track each item individually, it is less expensive. There is some advantage in simply maintaining an inventory of these items from which those who need them can draw; the purchasing department or stores personnel know, based on some established inventory level, to keep a set number of these items.

The advantages of buying and stocking may occur when a second device is used with or integral to the function of a bulk component. For example, a thermocouple or thermometer might be used with many different lengths and types of wells, and to stock them all would not be efficient. On the other hand, it might be preferable to have the factory assemble a diaphragm seal with a pressure gauge or transmitter so the fill fluid is properly installed in an experienced shop with all the required tools and QA (quality assurance) standards, rather than in the field. Technicians and contractors have far more important uses for their skills and, ultimately, field work is expensive.

Some projects or firms use a generic identification tag number for some uncomplicated devices. An example from ISA-5.1 uses PG-200 to designate a pressure gauge with a range of 0–200 psi. The PG-200 designation might appear 50 times on a complete project set of P&IDs. Another pressure gauge tag, PG-150, might also be used for gauges similar to PG-200 except with a different range, 0-150 psi. PG-150 might appear 25 times on the P&IDs. To prevent confusion, some firms use a letter designator in lieu of the loop number to ensure that these commodity (bulk) device tag numbers are not mistaken for a loop number.

Expanded Commodity Approach

It is interesting to expand this commodity approach to your storeroom stock or a construction stores trailer. With the advent of digital instrumentation that has a comparatively wide range of capability with factory calibration that will rarely drift, and simply requires digital range modification to meet a specific application, high dollar primary measurement devices like pressure transmitters can also be considered an off-the-shelf commodity. Draw the device from stock, range it per your specification data sheet and install it. Simple. The authors have seen this approach in a large industrial facility and on some construction sites: the purchasing department enters into an agreement with the instrument supplier, the supplier stocks the shelves and owns the product until someone draws the component. This has cost advantages for the company buying the device as well as locking in a customer for the company selling the device - it is a win/win.

Filling Out Specification Forms

ISA has developed a software version of ISA-20 containing all the forms and instructions on a single CD. It is possible to select a form, fill it out from a computer keyboard and print the finished Specification Form. The software version of ISA-20 has been used to develop the Specification Forms in this chapter.

In addition, the software version of Technical Report ISA-TR20.00.01 is available on a CD. The Specification Form for Displacer-Type Level Transmitters or Local Controllers is included as Appendix C. The pick list permits selecting the variables in a Specification Form and printing them on the Specification Form.

Figure 4-2 shows the Specification Form for pressure gauges from ISA-20, and Figure 4-3 shows instruction sheets to help fill out the form.

We will review the method used to fill out the pressure gauge Specification Form by following the instruction sheet.

1. A direct reading gauge is required.
2. Local mounting is required. The pressure gauges are mounted on a piping connection to a process line as shown in an installation detail. (See Chapter 8 Location Plans and Installation Details.)
3. Dial diameter 4-1/2". Dial color is white.
4. Case material – Manufacturers Standard Plastic. Phenolic is common, but other plastics are used; the point is to not rule out a supplier because they don't use the same plastic as the one you listed. This assumes that any plastic is acceptable for the atmosphere in which the gauge is located.

Figure 4-2: Pressure Gauge – Specification Form

ISA

	PRESSURE GAGES			SHEET	OF
				SPEC. NO.	REV.
NO	BY	DATE	REVISION	. 431	0
0	FAM	12/15/2003		CONTRACT	DATE
				1234	1/3/2003
				REQ. - P.O.	
				J-1	
				BY / CHK'D / APPR.	
				FAM / CAM / LF	

1. Type: Direct Rdg ☑ 3-15 lb Receiver ☐
   Other ______
2. Mounting: Surface ☐ Local ☑ Flush ☐
3. Dial: Diameter 4/12" Color WHITE 1/2%
4. Case: Cast Iron ☐ Aluminum ☐ Phenol ☐
   Other MFG. STD. PLASTIC
5. Ring: Screwed ☐ Hinged ☐ Slip ☐ Std ☑
   Other ______
6. Blow-Out Protection: None ☐ Back ☐ Disc ☑
   Solid Front ☐ Other ______
7. Lens: Glass ☑ Plastic ☐
8. Options: Sylphon Material ☐ ______
   Snubber ☐ ______
   Pressure Limit Valve ☐ ______
   Movement Damping ☐ ______
9. Nominal Accuracy Required 1/2% FULL SCALE

10. Mfr. & Model No. ______
11. Press. Element: Bourdon ☑ Bellows ☐
    Other ______
12. Element Mtl: Bronze ☐ Steel ☐ 316 SS
    Other ______
13. Socket Mtl: Bronze ☐ Steel ☐ 316 SS
    Other ______
14. Connection-NPT 1/4 in. ☐ 1/2 in. ☑ Other ______
    Bottom ☐ Back ☐
15. Movement: Bronze ☐ SS ☐ Nylon ☐
    Other MFG. STD.
16. Diaphragm Seal:
    Mfg. ______ Type ______
    Wetted Part Mtl. ______ Other Mtl. ______
    Fill Fluid ______
    Process Conn ______ Gage Conn ______

Rev.	Quan.	Tag Number	Range	Operating Pressure	Service
0	50	PG-150	0-150 psi	various	—
0	25	PG-200	0-200 psi	various	—

NOTES:

 ISA FORM S20.41a

From ISA-20

5. Ring – Standard. Let the gauge manufacturer specify the ring that holds the glass in the housing.
6. Blow Out Protection – This lets a part of the case blow out if the Bourdon tube ruptures. Since the operator will read the gauge from the front, a blowout disk in the back of the case should be specified.

7. Lens – Glass might break; plastic sometimes becomes cloudy. With a back blowout, glass has been specified. Shatterproof glass can be specified, glass similar to that used in a car windshield.

8. Options – No options specified.

9. Nominal accuracy required – +/- 1/2% is a good accuracy and does not keep manufacturers from quoting; also, it keeps the costs low.

10. MFG and Model # – To be filled in after gauges are purchased.

11. Pressure element – A Bourdon tube is usually used for the ranges under consideration.

Figure 4-3: Pressure Gauge – Instructions

21 Pressure gauges

Instructions for ISA Forms S20.41a and 20.41b

1) When receiver gauges are specified, the "Range" in the tabulation is the dial range.

2) Select mounting style.

3) Specify nominal dial diameter. Dial assumed white unless otherwise specified.

4) Select case material.

5) Specify ring style, or check "STD" if not important.

6) Specify blow-out protection. "Back" refers to a blow-out back. "Disc" refers to a blow-out disc located in the back or side of the case.

7) Specify lens material.

8) Options:

Snubber	Specify type or model number.
Sylphon Material	If sylphon required, specify material.
Movement Dampening	Specify if required.

9) Specify nominal accuracy, such as "±1/2%."

10) Write in make and model number after selection is made.

11) Specify element type or write in "MFR.STD."

12) If stainless steel is required, write in the type; such as "316."

13) See 12.

14) Specify connection size and location.

15) Specify movement or write in "MFR.STD."

16) If Diaphram Seal is required, fill in specifications.

For convenience, write in psig or other pressure unit at the top of "Range" and "Op. Press" columns, if all are the same.

From ISA-20

12. Element material – 316 stainless steel has been selected.
13. Socket MTL (material) – 316 stainless steel.
14. Connection – 1/2".
15. Movement – MFG-STD; let the manufacturer specify.
16. Diaphragm – None required.

Complete the form as follows:

Quantity	Tag	Range	Operating Pressure	Service
50	PG-150	0-150 psi	Various	—
25	PG-200	0-200 psi	Various	—

Then fill out the top right portion of the form—the revision number, sheet number, specification number, and contract number. You are now finished with the Specification Form.

One of the challenges you will face when filling out Specification Forms is that some options are mutually exclusive. For example, at least one manufacturer requires a plastic lens for filled gauges; glass is not available. So be sure you are addressing the real requirements and not the nice-to-haves. At some point any mutually-exclusive options will have to be reconciled.

Alternatives to the Use of Specification Forms

Some projects do not use Specification Forms to define measurement and control devices. These projects might define the devices using bills of material, requisitions, or a simple descriptive paragraph. These methods are not significantly different from using Specification Forms, since similar information is needed for all—complete definition of the device is necessary.

The decision to use Specification Forms, bills of material, requisitions or descriptive prose will probably be made by someone other than the control systems designer, but the decision should at least be influenced by the people using the information. Obviously, the Specification Form approach is better for the control systems designer since much of the work has already been standardized by ISA-20 or ISA-TR20.00.01.

If the maintenance group wants information for setting up their files, there is an advantage in using Specification Forms. If you or your designer is already set up to generate and maintain them, then there is an advantage to sticking with that approach. If your system just needs the devices purchased and no one else needs data in a set format, then buying the instruments using a text or prose description is fine. Today, there is a real benefit to maintaining whatever format you choose in an electronic form that can be emailed to the vendor for quote, forwarded to Purchasing for acquisition, or stored on a company's Intranet for reference by maintenance technicians.

If you are in a decision-making position, you may choose to call out instruments only by the manufacturer's name and model number instead of with a Specification Form. This approach limits the involvement by others on the design team since very few people understand the feature codes in a 10 or 20 character model number. Remember though, additional eyes are usually helpful with regard to the process data and materials compatibility. In addition, the model number approach implies that another document will have to be developed to establish the configuration and programming of the shared display – shared control devices including the DCS, PC, PLC and SCADA system designers.

This approach has at least three impacts on the purchase of the component. First, it cuts out the opportunity to take advantage of other instrument manufacturers' expertise, and thereby introduces more risk into the project; second, it gives an advantage to the named manufacturer and perhaps cuts out other possible suppliers, which may be the reason that this is done. Third, if the specification asks for an ABC Model 241B, that is what the vendor will supply, and you may miss advancements in their product line that may be advantageous to your application. In addition to this, the risk that the device may be incorrect for the application rests with you, the result of limiting the use of the expertise of the control systems community.

If you ask in a general way for a device that will bend widgets, many different suppliers' expertise will go into offering devices that will do that, and perhaps XYZ Company has a better, or perhaps, even a less expensive widget bender. Even if you are going to send the order to only one manufacturer, having all the data on a Specification Form will increase the likelihood that she or he might catch something you missed, or they could even come up with a new and improved model. There has been no small increase in the capabilities of instrumentation in the 20 years since you standardized on your widget bender.

Detailed Preparation of a Specification Form

The P&ID for our example project (see Figure 2-21) identifies all instruments by symbols and tag numbers. The next step for the control systems group is to prepare a Specification Form for each tag marked item. For demonstration, LT-100, the level transmitter in level loop LIC-100 (see Figure 2-24) will be the basis of the demonstration.

As a starting point, we will gather some information from the P&ID. LT-100 is an electronic displacement level transmitter, used to measure the level of 01-D-001, a 6' diameter vessel. ISA-20 contains a Specification Form for level instruments (displacer or float): ISA form S20.26. Please refer to Figure 4-4. The instructions for the form are included as Figures 4-5 and 4-6.

Figure 4-4: Level Instrument – Specification Form

LEVEL INSTRUMENTS (DISPLACER OR FLOAT)				SHEET	OF
NO	BY	DATE	REVISION	SPEC. NO. 321	REV. 0
0	FAM	12/15/2003		CONTRACT 1234	DATE 1/3/2003
				REQ. - P.O. J-6 J-12	
				BY FAM	CHK'D CHK CAM / APPR. LF

	1	Tag Number		LT-100		
	2	Service		K.O. DRUM		
	3	Line Number / Vessel Number		01-D-001		
BODY/CAGE	4	Body or Cage Material		C.S.		
		Rating		300 psi		
	5	Conn Size & Location Upper		1 1/2" TOP		
		Type		300 psi FLG		
	6	Conn Size & Location Lower		1 1/2" BTM		
		Type		300 psi FLG		
	7	Case Mounting		SIDE		
		Type				
	8	Rotatable Head		NOT REQ		
	9					
	10	Orientation		LEFT HAND		
	11	Cooling Extension		NOT REQ		
	12					
DISPLACER OR FLOAT	13	Dimensions		48"		
	14	Insertion Depth				
	15	Displacer Extension				
	16	Disp. or Float Material		304 S.S.		
	17	Displacer Spring/Tube Mtl.		MFG. STD.		
	18					
	19					
XMTR/CONT.	20	Function		TRANSMITTER		
	21	Output		4-20 mAdc		
	22	Control Modes				
	23	Differential				
	24	Output Action: Level Rise		INCREASE		
	25	Mounting		INTEGRAL		
	26	Enclosure Class		NEMA 8		
	27	Elec. Power or Air Supply		24Vdc from shared		
	28			display		
SERVICE	29	Upper Liquid		WET GAS		
	30	Lower Liquid		DEGASSED MTL.		
	31	Sp. Gr.: Upper	Sp. Gr.: Lower		.9 @ 60 F	
	32	Press. Max.	Normal	50 PSI	4 PSI	
	33	Temp. Max.	Normal	400 F	90-150 F	
	34					
	35					
OPTIONS	36	Airset	Supply Gage			
	37	Gage Glass Connections				
	38	Gage Glass Model No.				
	39	Contact: No.	Contact: Form			
	40	Contact Rating				
	41	Action of Contacts				
	42					
	43					
	44					
	45					
	46	Manufacturer		LATER		
	47	Model Number		LATER		
	48					

NOTES:

© 1981 ISA ISA FORM S20.26

From ISA - 20

Figure 4-5: Level Instrument – Instructions, Part 1

16 Level instruments (displacer or float)

Instructions for ISA Form S20.26.

1) Tag No. or other identification.
2) Process service.
3) Line number or vessel number on which cage or body is installed.
4) Material of chamber and/or mounting flange.
5) For float specify top or side of vessel connection. For displacer in a chamber specify upper, then lower connection; such as side-side, side-bottom, top-bottom, etc. Give flange size and rating or NPT size.
6) Same as 5.
7) Refers to position of case when viewing the front of the case relative to the chamber; the case is either to the left, right, or top.
8) On displacer instruments specify if case is to be rotatable with respect to the chamber. This only applies if there is one or more side connections.
10) Orientation of control with respect to displacer cage.
11) Cooling Extension.
13) Specify float diameter or displacer length. The displacer length is also the range.
14) Insertion depth applied to ball floats. It is the mounting flange to the center of the ball.
15) The displacer extension is measured from the face of the mounting flange to the top of the displacer. This dimension is required only for top of vessel mounted instruments.
16) Includes rod.
17) Refer to MFR's standard materials or special materials.
20) Transmitter, controller, switch, etc.
21) Air pressure or electrical signal output of transmitter or controller.
22) P: Proportional

    Pn: Narrow band proportional

    PI: Proportional plus Integral (Reset).
23) Differential if controller on/off must specify differential adj. or fixed. State adjustable range or fixed amount.
24) INCREASE (Direct action) or DECREASE (Reverse Action).
25) Remote, or integral.
26) Electrical classification of housing. NEMA number.
27) Air pressure or voltage. If electronic, state whether ac or dc.

From ISA-20

Figure 4-6: Level Instrument – Instructions, Part 2

30) Used for all services.

31) Specific gravities at operating temperature.

32) Operating and max. pressure, or vacuum.

33) For cryogenic service, give minimum temperature.

36) Airset assumed mounted to case.

37) Connections on chamber, give size.

38) Specify gauge glass, if required.

39) Contact form: SPST, SPDT, etc.

40) Give Volts, Amps.

41) Describe contact action with level.

47) Model number of entire assembly.

From ISA-20

Starting at the top of the form, we fill it out as follows:

1. Tag number – LT-100
2. Service – Knock Out Drum
3. Line No./Vessel No. – 01-D-001

Body/Cage

4. Body or cage material – Carbon Steel

   Rating – ANSI 300 (information from the piping group)
5. Connection Size & Location-Upper – 1 1/2" top (coordinate size with piping group)

   Type – ANSI 300 raised face flange
6. Connection Size & Location-Lower – 1 1/2" bottom (coordinate size with piping group)

   Type – ANSI 300 raised face flange
7. Case Mounting – Side
8. Rotatable Head – Not required
9. This space is blank on the form and also in the instructions, so it can be used as you wish or left blank.
10. Orientation – Left Hand

    Some instruments are fabricated with the scale and the access for maintenance available at only one point so orientation of the instrument is important. Orientation is coordinated between the vessel or piping designers and the instrument specialist. Typically, the specialist will use the instrument supplier's orientation charts to ensure that the piping group understands the clearances and orientations offered. This is an area of frequent misunderstanding so attention to detail is warranted.

11. Cooling Extension – Not required
12. Blank, same as 9, above

Displacer or Float

13. Dimensions – 48"

    Note: We will measure only the liquid height in the center 48 inches of the tank diameter.
14. Insertion depth – Does not apply
15. Displacer extension – Does not apply
16. Displacer or float material – 304 stainless steel
17. Displacer spring/tube material – MFG STD

XMTR/Controller

20. Function – Transmitter
21. Output – 4 to 20 mA DC
22. Control modes – Does not apply, control is remote in a DCS or PLC
23. Differential – Does not apply
24. Output action level rise – Increase
25. Mounting – Integral
26. Enclosure class – NEMA 8
27. Electric power or air supply – 24 VDC from a DCS or PLC

Service

29. Upper liquid – Wet gas
30. Lower liquid – Degassed material
31. Sp.Gr. – Lower .9 @ 60°F.
32. Press Max / Normal – 50 psi / 4 psi
33. Temp Max / Normal – 400°F. / 90°–150°F

Options 36 – 45: not required.

46. Manufacturer – Later, after device is purchased
47. Model Number – Later, after device is purchased

Then, we fill out the upper right portion of the form: the revision number, sheet number, specification number, and contract number. The form is now complete.

Most projects have or develop overall specifications, called master specifications. These specifications might be specific for control systems, or they may be much broader and include all parts of the project. They may be a few pages, or they may be a complete series of specifications for all items supplied for the project. They may be printed and bound into many volumes. If you are filling out Specification Forms you should refer to the master specifications to ensure that the devices you specify comply with them.

Classified Production Areas

Many processes involve handling flammable materials and most use electricity in one form or another. Flammable material, when combined with electrical equipment, becomes a potential hazard. Figure 4-7's third bullet introduces the way this potential hazard may be reduced.

Figure 4-7: Hazardous (Classified) Locations

- Chemical and petrochemical plants, petroleum refineries and other installations which use materials that are flammable
- Many control system devices use electricity
- To prevent fires or explosions, equipment use and installation must follow safe practices that are set down in codes
- Information on these codes follows...

The installation of all electrical and electronic equipment must follow rigid rules set forth in the National Electric Code® (NEC®). The NEC is the law in the United States. Similar, but different codes exist in Canada, Europe and elsewhere. In the United States, the craftsmanship of licensed electricians and their knowledge of the code ensures that work is done in compliance with the NEC, and local building or electrical inspectors verify compliance with the law. It is important that those designing a project understand the roles and responsibilities of code compliance.

You should refer to the Code and the year in effect in your jurisdiction before doing any design or other work, but elements of the document are summarized below.

The extent of the potential hazard is mandated by the NEC in a classification system consisting of three parts: Class, Group, and Division (See Figure 4-8).

Figure 4-8: Area Classification – Electrical

- For plants where flammable materials are present, the National Electrical Code (NEC) has set up a classification system consisting of three parts:

  CLASS

  GROUP

  DIVISION

  – For example, Class I, Group D, Division 1

The **Class** designation denotes the generic nature of flammable materials.

Class I locations – where flammable gases or vapors may be present in the air in quantities sufficient to produce an explosive or ignitable mixture (e.g., chemical and petrochemical facilities).

Class II locations – where combustible dusts may be present in sufficient quantities to cause hazard (e.g., flour mills and coal pulverizing facilities).

Class III locations – where the hazardous material consists of easily ignitable flyings or fibers that are not normally in suspension in the air in quantities to produce ignitable mixtures (e.g., sawmills, textile mills).

The Group designation defines the hazardous material, for example:

Group A – acetylene

Group B – butadiene, hydrogen

Group C – carbon monoxide, hydrogen sulfide

Group D – ammonia, most petroleum products

Group E – metal dusts

Group F – carbon black

Group G – flour

Note: This list is only an example; the complete list is much more comprehensive. The NEC also specifies temperature-rating code, known as the "T" number.

The **Division** designation specifies the frequency of the hazard, for example:

Division 1

- Hazardous concentrations normally exist or often exist
- Frequent repair or maintenance causes hazardous concentrations to exist

A **Division 1** location is one where:

- Hazardous concentrations exist continuously, intermittently or periodically under normal operating conditions
- Hazardous concentrations exist frequently because of repair or maintenance operations or leakage of equipment
- Breakdown of equipment or process failure might simultaneously release hazardous concentrations of flammable gases, vapors, or dust, and cause failure of electrical equipment

Division 2

- Hazardous only in abnormal situations
- Hazardous materials are handled, but normally confined
- Nonhazardous areas because of forced ventilation
- Dust layers may accumulate

A **Division 2** location is presumed to be hazardous only in abnormal situations, such as the result of an accident when process equipment or a container fails. A Division 2 location is one in which:

- Flammable liquids or gases normally confined in closed containers or systems are handled or processed
- Areas normally not hazardous because of forced ventilation become hazardous if ventilation equipment failed
- Areas adjacent to Division 1 areas, where hazardous concentrations could be communicated unless prevented by positive ventilation with adequate safeguards
- Areas where layers of hazardous dust can accumulate

An area classification plan is developed, usually by the electrical designer of record, who defines the class, group and division of all areas of a facility. The wiring must follow NEC practices set forth for that area's classification. Of particular importance in this chapter is that all equipment installed in the hazardous areas must be labeled by the manufacturer as being suitable for that area. The device must carry the label, usually imprinted on the manufacturer's permanent identification label.

The area classification for an instrument is furnished to the vendor by a nationally recognized laboratory under criteria furnished by the labeling authority. The largest and most accepted labeling authorities in the United States are Underwriters Laboratories (UL) and Factory Mutual (FM). The Canadian Standards Association (CSA) label is also frequently accepted in many jurisdictions in the United States. The CE mark from Europe is not normally accepted in the United States, but many instruments carry all four labels.

The label lists the classifications for which the device is rated, and that must match the classification of the area in which it will be installed—for example, Class I, Group D, Division 1. Also available in the vendor's literature for the device, and possibly on the label, is the National Electrical Manufacturers Association (NEMA) type as defined in NEMA Standard 250-2008. The NEMA type applies to the enclosure of the device. Figure 4-9 describes the NEMA types of enclosures for hazardous locations.

Figure 4-9: NEMA Standard 250-2008

NEMA enclosure types for hazardous locations		
Type	**Indoor/Outdoor**	**Suitable For**
7	Indoor	Class I, Div. 1, Gr A,B,C,D
8	Both	Class I, Div. 1, Gr A,B,C,D
9	Indoor	Class II, Div. 1, Gr E, F, G
10	–	Mine Safety and Health Administration, 30CFR, Part 18

- See NEMA 250-2008 Enclosures For Electrical Equipment (1,000 Volts Maximum) for complete information.

Intrinsic Safety

Many instruments require very little power to operate. They can be installed in Division 1 hazardous locations using Intrinsic Safety techniques when properly labeled.

Figure 4-10 describes the concept of Intrinsic Safety. Equipment and apparatus that are incapable of storing and releasing enough energy to cause ignition of the hazardous atmosphere present must be used. This includes specially designed devices (electronic transmitters, for example), Intrinsic Safety barriers and a wire and cable system with low capacitance and inductance.

Figure 4-10: Intrinsic Safety

- **Basic premise**
  - It is possible to construct a circuit that is incapable of storing and releasing enough energy to cause ignition of a hazardous atmosphere under normal, abnormal, or fault conditions
- **Implementation**
  - The energy supply to the intrinsically safe circuit is limited by a barrier
  - Energy storing components are prevented from storing too much energy
  - Or, the devices in the circuit are proven to be inherently, intrinsically safe and are employed in a properly designed intrinsically safe system

ISA-TR12.2-1995 *Intrinsic Safety System Assessment Using the Entity Concept* contains further detailed information on Intrinsic Safety.

Further discussion of area classification and Intrinsic Safety is beyond the purview of this book. There are many training opportunities, books and equipment supplier based training that you can use to enhance your knowledge.

CHAPTER FIVE

Purchasing

Overview

Occasionally some readers may be involved with the purchase of the numerous components that make up a control system. Just as there is terminology, documentation, procedures and traditions in control systems, there are these same things with the people that acquire your equipment. If you develop an understanding of their rules, traditions and language, your coordination will be better and the process will be smoother. In this chapter we will proceed through a typical purchasing process; yours will likely be different in the details, but there will be similarities in the procedures. Let us assume that the project design phase is complete and the stack of Specification Forms, or data sheets, is complete and issued "For Purchase" or "Construction". It's now time to buy the devices. A typical acquisition process might consist of the seven steps shown in Figure 5-1.

Figure 5-1: Seven Steps of Acquisition

1. Develop an acceptable suppliers list
2. Assemble the bid packages
3. Send the bid packages out for bid
4. Formally receive the proposals
5. Evaluate the proposals
6. Award the purchase order
7. Receive the devices

We use the term purchasing, purchasing group and purchasing professionals throughout this book to identify the people in your organization that handle the commercial and even legal aspects of this effort. You might refer to them as "stores" or procurement or buyers. While the terms are not really interchangeable within their profession, we are looking at this from a less precise aspect and we use the terms as if they were interchangeable, with apologies to them in advance.

Many control system practitioners will not need to personally perform all the listed steps, but someone will. We explain the steps so you have an understanding of what is going on "behind the scenes"; this might even explain some frustrating time delays that you have experienced. You might be lucky enough to be able to draw the instrument or final control element from the storeroom without having to consider how it came to be there. Other control system practitioners will have been intimately involved in the purchasing process and will have selected acceptable suppliers, solicited bids, evaluated the bids, selected the successful bidder, and requisitioned and purchased components. For some smaller companies, you may even issue purchase orders under the direction of a purchasing professional and maintain receiving documentation, although these situations are rare.

Refer to the list in Figure 5-1, Seven Steps of Acquisition. At your place of employment, there may be no need for this lengthy evaluation process. However, some day you may work on a project where all this has to be formally done—hence, the following discussion.

Seven Steps of Acquisition

Figure 5-1, Seven Steps of Acquisition, is an outline of the process necessary to get an l instrument to your facility. The process involves at least three separate work groups as well as the instrument supplier: (1) control system designers, (2) the purchasing group, and (3) owner representatives, including maintenance and management. All of these work groups may have definite opinions about who the suppliers should be and the purchasing procedures that should be followed.

A control system designer has the experience to develop a list of acceptable suppliers. These suppliers will have an understanding of advances in the industry and in the latest capabilities of control system devices. The control system design group should have some understanding of unresolved problems your facility has had with the technical performance of a particular manufacturer in the recent past or with the service that manufacturer has provided. It is appropriate to use your experience to modify the acceptable suppliers list. The purchasing professionals may perform checks on the commercial viability of a manufacturer to determine if it is appropriate to enter into a business relationship with that company.

In your organization, the control system designer may not see priced bids during the evaluation process. The concept here is that the technical evaluator should focus only on the technical features of the bid. This approach may be less common today since everyone should be interested in getting the best value for the money expended. However, it is generally true the control system designer wants the highest possible quality and the best features for their devices. Understand though that features cost money, and complexity means additional training and possibly confusion for maintenance activities when the device fails a year from now. There really should be a clear need for the accuracy you choose, the precision, multiple sensor input capability, materials of construction and a myriad of other choices available before you buy or specify them. On the other hand, there is also significant value in having robust devices that withstand the environment and process deviations without drift.

The purchasing group looks at the business record of each supplier, and asks hard questions: "Does this supplier have the capability to deliver the devices needed in the time available, manufacture the quality required, and provide the necessary technical and administrative support?" In addition to questions of quality and support, the purchasing group wants to ensure that the components purchased will actually be delivered and will arrive within some semblance of the schedule. They manage the risk of doing business with the potential supplier.

Owner representatives may have had both good and bad experiences with some of the potential suppliers. If a project is an expansion of an existing facility, the owner representatives may have definite ideas as to whom the suppliers ought to be.

Considerable diplomatic skill and a cooperative spirit among the three groups are often necessary to proceed through the Seven Steps of Acquisition in a timely manner.

Step 1: Develop an "acceptable suppliers" list.

On our example project, we are going to ask for competitive bids on supplying a group of control system devices. We are assuming that no supplier pre-selection has been made. Preparing an acceptable list of suppliers is a bit more difficult than it probably sounds.

Pre-selection is where your company performs many of these acquisition steps in advance. When you have pre-selected a manufacturer, you are able to simply send your data sheet to the manufacturer and have it shipped. The bidding, evaluation and pricing is already done. These pre-selection agreements involve significant negotiation and effort to develop effective pricing, in advance, for a range of devices. The manufacturer will probably want some quantity commitments from the owner so they can establish pricing; in general the more you buy the lower the price. Pre-selection is good for the facility owner because it becomes quite simple to "buy" devices, all the time intensive work is already complete. It is better for the manufacturer because they now have an official relationship with the facility owner and they have, by virtue of the quantity commitment, a good idea of what will need to supply over time, and they don't have to expend as much effort in continuously bidding on new lists.

More commonly, three or more suppliers are needed for an "acceptable suppliers" list. Three suppliers seem to adequately meet the goal of truly competitive pricing. Yet, contrarily, a minimum number of bid packages are desirable, since preparing and handling bid packages takes time and costs money. As a control system professional, you understand the market in your area and in your industry. You know the instrument supplier representatives in your area. You know which suppliers have experience in your industry. Suppliers really shouldn't be on the list if they don't have the knowledge and experience to handle the special challenges of your industry or facility. For example, your process may call for devices to be cleaned for oxygen service, you may need to have sanitary connections on your instruments, or your industry may have standardized on special surface treatments and cleaning requirements for ultrapure water systems or possibly special diaphragm seals for paper pulp stock or wastewater applications.

In addition to having the knowledge and experience to handle the special challenges of your industry or facility, what are the criteria for being an acceptable supplier? If it is an existing plant, the owner's experience is invaluable (and is likely to be a deciding factor). Today, the differences in the capabilities and performance of control system devices furnished by different suppliers might actually be negligible (although the manufacturers may not agree with that statement). So, what sets suppliers apart? Investigation and caution are appropriate if your maintenance or engineering staff has never seen a particular supplier at your facility. If training is required for the system or devices, the cost of conducting that should be factored into the selection process, particularly when the training site is remote. Casual or "I just dropped by, how is it going" support can be quite effective, particularly for a new system or complex piece of equipment; this also should be a factor in your selection process. Personal service is an important feature not every supplier can, or will, furnish. If your company has no history with one particular supplier, but a good history with other suppliers, you might want to think twice about including the unknown company on the acceptable suppliers list. And don't neglect to include hidden costs in your evaluation and when setting up your bid documents; is there a software fee or is there a yearly license fee to support the product? Are special cables or reagents needed and what is their cost, now and in the future, or interestingly, what were the fees over the past five years?

In some industries and areas of the country, it is not uncommon for instrument suppliers to check in with maintenance technicians monthly or even weekly, or for manufacturers' representatives to host classes in applications and software configurations, or to make impromptu presentations on advances in features. If you have an installed instrument base at your facility, the suppliers should check in from time to time to see how their devices are performing. The importance of local support is obvious; include it as a part of your bid documents and evaluation!

The project's standard request for quotation (RFQ) or bid letter might include a statement that says your company reserves the right to award the purchase order to anyone—meaning that the low bidder will not necessarily receive the award. This provides significant latitude in selecting the successful supplier. However, all bidders should have an even chance of success. If you are "only going through the motions" knowing full well you will award the order to "Brand X," you are doing a disservice to the other bidders. If the decision to go to a particular supplier has already been made, it is more efficient and honest to work with your purchasing department to negotiate with the supplier of choice, to develop a purchasing agreement, without going through the bid process. It is a terrific waste of time, effort and good will to require bids from suppliers while knowing they will never be successful. That being said, if you are willing to change suppliers at your facility, your company can occasionally reap some savings during competitive bidding. We encourage you to involve the maintenance staff in any decision to change device manufacturers; your willingness might not be shared by the rest of your team. Not everyone likes change.

Many, if not most, facilities standardize on a single instrument manufacturer for a particular type of control system device within a facility. This simplifies purchase, installation, training and the stocking of spares. Clearly, when you have one family of magnetic flowmeters, your maintenance group will be familiar with configuring and troubleshooting the devices. The face-to-face dimensions will be the same when replacement is required, and the wires will all connect to the same component in the same place. Your installers will know the proper way to mount and ground them. Standardizing on a single supplier may also allow your facility's management to enter into special, reduced pricing agreements with the supplier.

Step 2: Assemble the bid packages.

A typical list of bid packages for a reasonably sized project might include the following groupings. The list is for illustration only; the actual breakdown will change to meet the specific requirements:

- Pressure transmitters, including pressure, differential pressure, d/p flow, and flange-mounted level transmitters
- Magnetic flowmeters
- Vortex flowmeters
- Turbine flowmeters
- Ultrasonic level transmitters
- Temperature sensing elements
- On-off control valves
- Modulating control valves—globe and ball
- Modulating control valves—butterfly
- Control panels
- Distributed Control Systems

For effective bidding, individual packages should only group those devices normally furnished by the listed bidders. That is, don't package pressure transmitters with mag meters (magnetic flowmeters) in one bid package unless all the bidders on the list normally furnish both types of devices. On the other hand, if all your proposed suppliers have, in their product line, magnetic flowmeters, and pressure transmitters and, for argument's sake, vortex flowmeters in their product line, then go ahead and make one package for all the devices. Remember that some suppliers will not submit a bid if they can't furnish the complete package, so be sure of their scope and capabilities.

The Bid Package

The bid package is the group of documents sent to potential suppliers to guide them in preparing a proposal to furnish a product. In its simplest form, the bid package might be a faxed Specification Form. In its most complex form, a bid package might be 27 volumes of design documents and specifications used to design and build a billion-dollar refinery. For purchasing instrumentation, a bid package probably contains a cover letter explaining what you want from the supplier, called a request for quotation (RFQ). This letter will contain:

- Contact names, address and phone number for the company that will purchase the devices
- The name and address for the entity that pays the invoice – the "Bill To" address
- Delivery location, called the "Ship To" address
- Schedule: states when the devices are needed
- Bid data: Bid due date, what documents should be included with the bid, estimated duration of the evaluation process. Documents needed with a bid should be listed and may include requests for warranty statements, installation, operation and maintenance manuals, technical bulletins and dimensional drawings. Note that "Certified" drawings needed by the piping designers are not normally furnished until after the purchase order is issued.

Finally, with all the organizational information out of the way, the bid package should, of course, include the ubiquitous Specification Forms.

In addition to the cover letter, a bid package includes the following components:

1. Invitation to Bid

The Invitation to Bid may be part of the cover letter, or possibly a separate document furnished by project management or by the purchasing team to cover this standard information:

a. Who is requesting the bid and who will be buying the components? The supplier should know who to contact for questions. They will also want to establish the specific company that will purchase the devices; whose "paper" will be used. "Paper is a slang term for the formal documents that make up the legal purchase document package, the purchase order and the terms and conditions, as a minimum. This becomes important for projects when contractors are purchasing control system devices since the manufacturer is not selling the devices to the ultimate owner, the operating facility; they are selling to the construction contractor. This can have warrantee implications that the purchasing professionals will work out with the suppliers and the contractor. The supplier will do a credit check on the purchaser, the contractor, rather than on the ultimate owner, your facility, just as the purchaser will do one on the supplier. An unfavorable credit history may have an impact on pricing.

b. What is the project name? Suppliers will want to know if the quote is for ongoing work at the facility or if the quote is for a "grass roots" project. Sometimes, suppliers have access to special pricing for project work, so this could work to your advantage.

c. Where will the devices be used? The ultimate destination or, simply the installation point, should be defined. The purchase orders used for your project might not clearly show the location of your facility, particularly when the work is designed remotely from your site and purchased and installed by an at-the-home office of a contractor 1000 miles away from your facility. This is important because some equipment suppliers portion out credit and profit for a sale within their organization to the representative supporting the design effort and to the representative supporting construction and operation. The local representative that supports the operation of your facility will be happier if they get paid by the manufacturer for the services they furnish. While this is not really your responsibility, it is nonetheless better for you if everyone is included in your team. It is helpful to define the players up front.

d. Dates: When is the bid due—also known as the Closing Date? When do you expect to evaluate the bids? When will you award the order? When do you expect to take delivery? And, when do you expect to install the product? For larger bid requests, the supplier should be given from two weeks to a month to prepare the quotation. Be sure to take into account mail delivery times for sending and receiving the bid package. Simple quotes can be prepared in a day or so, and, today, more and more quotations are delivered using email. Be aware that some less technologically savvy organizations still only allow facsimile transmissions of bids, but that too is evolving. The various dates all have an impact on the suppliers' quotation workloads and on their production schedules. The device suppliers might need to block out a production window to meet your requirements; they need to know delivery requirements up front.

e. The letter should clearly state what you want back from the bidder. At the very least, price and delivery are needed, but bid packages may include a quotation sheet that requests a statement of compliance from the bidder on specification items and commercial requirements. Occasionally there is a request to provide an un-priced quotation and a priced one, used when the technical bid evaluation is separate from the commercial evaluation. You may want specification bulletins, installation, operation and maintenance manuals, or dimensional drawings to be included with the bid to facilitate quotation evaluations. Sometimes, orifice plate calculations and valve calculations performed by the device supplier can be included with the bid. All your expectations should be called out in the RFQ.

f. The use of electronic media. The format used to submit technical information with the bid and in support of the order for the successful bidder needs to be defined in the bid documents. Paper copies of bulletins and manuals are becoming rare. Operation and Maintenance documentation, and the storage of that information, are rapidly transitioning to paperless. The equipment and instrumentation suppliers should know what your media requirements will be. That being said, they will undoubtedly welcome the move to electronic media.

2. Specification Forms

   a. Include a table of contents, so the bidder knows they have a complete set of forms for bid. It is not unheard of for Specification Forms to get lost in transport.

   b. Check that the copies of the Specification Forms are legible. Confirm that the data presented is sufficient to define the requirements; preferably use the ISA format described in Chapter 4. A completed ISA Specification Form will provide sufficient information to bid the component. Check that erroneous information wasn't accidentally included as the forms were developed; it is disturbingly easy to neglect to clear out old data when generating new Specification Forms from older electronic files. Check for agreement between the Specification Forms and the text specifications. Check that the specifications provided to the device supplier are complete and appropriate for their responsibilities. For example, the authors have even seen construction installation specifications added to instrument bid requests, which prompted calls from the bidders asking if they were to be responsible for the physical installation as well!

Specification Forms may be prepared in at least the following three ways:

1. *Generic*. The designer distills down the control system devices that are likely to be furnished so comparative pricing can be developed. The Specification Forms are based on expected types and typical ranges of devices rather than on specific applications. They are sufficiently complete that vendors may offer firm quotations for types of devices. Generic forms may be seen when competitive bidding is used to pre-select instrument manufacturers. The pre-selection might take place before all

process requirements or the device counts are known. This generic method works a bit better when you provide a reasonably accurate estimate for the expected quantity of the components. You should ensure the bidders know the quantities are estimates and that specific tag number-based Specification Forms will follow with the actual order.

2. *Partially complete.* The Specification Forms are prepared with tag numbers, application-specific technical data and project requirements, but without the manufacturer's name and the device model number. Potential vendors complete the form with information specific to their devices; at minimum they list the complete ordering model number. The information supplied by the successful bidder is used to complete the Specification Form for purchase and for record. This is probably the more common approach. Care should be taken to list the required features and to leave a bit ambiguous or generic the features that are less important. An example is fill fluid, used in mechanical gauges to steady the indicating needle. For a benign process, when the supplier's standard (as in "no additional charge") fill fluid is adequate, it can cause problems if you list the proprietary fill fluid used by one supplier. There have been cases where well-meaning but less-informed people have tried to get an instrument supplier to use the specified fill fluid only offered by a competitor. Time, effort and goodwill are wasted while such problems are resolved. By addressing the wording used, maybe by adding "or equal," problems like this can be prevented.

3. *Manufacturer-specific.* Manufacturer-specific Specification Forms contain all the information of the partially complete form, but they also include component-specific information from the manufacturer, including the manufacturer's name and model number. This approach can be used to better define the device intended for an application without having to list each and every one of the numerous options and features: "Gimme one of those." It also unambiguously tells the other bidders a lot about their competition, along with the features the company requesting the bid finds important. If a manufacturer-specific specification form is used to define the features of an acceptable device with the intent to open the bid to all approved manufacturers, care should be taken to make it clear to all bidders that the listed manufacturer and model is one example of many that will be acceptable. You want to emphasize your intent that the bidders bid on an "or-equal" basis and that they describe any differences between the specifications and their products. The purchasing professionals can assist with this.

Choosing what kind of bidding you want to use is really a function of your purchasing process, your experiences and the design team. Generic or partially complete Specification Forms are prepared when there is concern that listing a device manufacturer will give that supplier an advantage over others. More than one potential supplier has declined to bid when seeing their competitor's name, or even features pointing to that competitor, listed on the Specification Form. It takes time, and therefore money, to prepare a bid. If the potential supplier feels that a bid will have a low probability for success, that supplier may very well decline to expend the effort. You might end up with only one, not very aggressively priced bid. It is relatively easy to reassure the bidders that their offering will be looked at in detail. The advantage of having the design built around a complete specification with complete model numbers is that the designer will work out the metallurgy, the wetted parts, the ranges, the capabilities, and all the other features before the component goes out to bid. Generic specifications tend to miss important details than can be onerous to work out during bidding, in our opinion.

On the other hand, there may be a hidden advantage to generic Specification Forms when competitive bidding is used. The design professional has more opportunity to concentrate on the process information, materials and required performance of the instrument instead of expending design effort figuring out the manufacturer's model number codes and options. The more effort expended

Tagging

Do you buy your instrument identification tag as part of the device model code? Many suppliers include that option. Many specification forms include a standard note that calls for the device supplier to "Permanently affix a stainless steel (SS) tag engraved with the device tag number given above," or some similar phrase. The device supplier dutifully checks a box on their factory order sheet and somewhere down the line a tag is added to the device or one is tossed into the shipping container. Consequently, the facility ends up with as many different styles of instrument tags as there are suppliers of devices.

Sometimes, the facility ends up with MORE tag styles than instrument manufacturers—the factory tag is different from the tag furnished by the local supply house! Occasionally, but assuredly, instruments will evade the process altogether and will arrive on site without any tag or with an incorrect tag. The furnished tag will be tossed out with the shipping box, or it will magically fall off the correct device and get reaffixed to the wrong device later. There must be a better way.

Remember, in most cases the separate tag that was requested on the Specification Form is an option from the device supplier, so there is an associated cost. We say "separate" tag because many instruments and control valves already include a nameplate riveted to the device, engraved by the manufacturer with the manufacturer's logo, the model number, the serial number, and the owner-furnished tag number. These nameplates may not be particularly easy to read in the field after some months of operation, but they are perfectly legible for warehousing, calibration and construction.

The cost for the optional separate tag can be significant—$25 and up in some cases. For your money, you end up with all types of tags, as well. In the authors' experience, these have ranged from an engraved phenolic plastic plate to a piece of thin SS tape hand scratched with the tag number. Furnishing your site tag specification to an instrument manufacturer is rarely successful or inexpensive. They are in the business of manufacturing high quality instruments, not tags.

It makes a lot more sense for the end user to specify and purchase all instrument tags at one time, to one specification, and from one supplier. This way the end user can control the material, engraving method, font size and attachment method. The tags are sure to meet the specifications because one entity furnished them. The authors have seen successful systems where the commissioning or startup team affixed tags as part of their procedures—the presence of a tag flagged a device as ready for operation.

The cost of an engraved SS plate with heavy-duty SS wire can be a fraction of the tagging option cost from the instrument supplier. Another successful tagging option uses plastic plates with a printed adhesive bar code in addition to the bold tag printing. The bar code quickly links the physical device to the site database for review, or specification data and calibration history, and it simplifies maintenance record keeping.

determining the minimum, normal and maximum conditions the device will operate under, the fewer the surprises during operation and the better the system. The minimums and maximums that make up operating ranges are important!

Competitive bidding on instruments may be falling out of favor. Many corporations have partnered with their control system suppliers, so purchasing new devices may simply mean calling the supplier on the phone and invoking a corporate buying agreement. For facilities that have a history with particular suppliers, the purchasing department may just negotiate a pricing structure based upon historical data and goodwill.

Step 3: Send the bid packages.

Bundle up the bid packages and send them out, or forward them to the purchasing department for handling. Many companies severely limit contact with potential suppliers during the selection process so that there is no possibility of unfair advantage given to one bidder over another. Do yourself a favor and follow company rules.

It sounds easy, and it is, to prepare a bid package for distribution. The following are some suggestions for a smooth bidding process based on past "opportunities," or, in other words, really bad experiences the authors have had over the years.

a. Verify the bidder's name and address. The business card you used for the address might be several years old, or perhaps the salesperson that calls on you spends all of his or her time on the road and rarely picks up their mail. The bid package may need to go somewhere other than the mail drop address on the card.

b. Request the purchasing group to notify the recipients that the package is coming. Ask them to then wait a few days and call again to confirm its receipt. More than one bid was never received by a potential supplier. Mail sent is not necessarily mail received. This is even truer with email; many host and recipient sites limit the size of files sent, or the attachments might have been stripped off the email. The clock is ticking, but the person who is supposed to be preparing the bid isn't.

Step 4: Formally receive the quotations (bids).

The professional way to handle quotations is for someone, often the buyer or purchasing agent, to hold all the bids unopened until the date and time mentioned on the Invitation to Bid. At the stated time, the packages are opened all at once, dated, and distributed for evaluation according to the company's rules.

The evaluation process can be an activity for three groups—purchasing, the design team, and, possibly, the owner. The purchasing group may perform a formal bid opening with witnesses at the stated time of bid closing. The goal is to give no bidder an advantage over another, to offer no pricing hints to the other bidders, nor to give any bidders grounds to mistrust the company's impartiality. Contents and pricing are confidential, except in the public sector, and should not be shared with the other bidders. The bidders may call you, as is seems natural, "How did I do?" It would be unfair to the other bidders to respond; forward all such calls to the purchasing department so there is no appearance of impropriety.

Step 5: Evaluate the Bids.

There are two aspects to a bid evaluation—commercial and technical. The design team does the technical evaluation. The owner might contribute to the evaluation as they have a stake in the features of the devices purchased. The purchasing department will probably do the commercial evaluation. They will address the warranty offered by the seller, the FOB call-out (the transfer of ownership during shipping), and other points included in the Terms and Conditions section.

In the technical evaluation, the instrument proposed is compared against the specification. Does it meet the requirements set forth on the Specification Form? A spreadsheet is a great tool for making this comparison. On the left column, the specification items are listed from the Specification Form: the performance, metallurgy, connection size and method, accuracy and repeatability, range, and anything else of importance.

In each subsequent column, the specific offering from each bidder is listed, with the evaluators' determination of whether or not the offering meets the specification. When the evaluation is complete, the columns are complete. It is not uncommon for all of the bids to meet the specification requirements. In this case, the technical evaluation would state that the bids are equal. Then the selection will be made on purely commercial terms.

As has been emphasized, care must be taken to evaluate against the Specification Form—not against some attractive, but unnecessary feature offered by one bidder that exceeds the requirements. Necessary features are specification items; unnecessary features can be, well, expensive fluff.

Shipping

The cost of shipping, or freight, for an instrument can be a significant, if somewhat hidden charge. It will be part of the commercial evaluation. It is not uncommon for shipping to be 15% of the total cost of an instrument, or more. Understanding shipping costs allows the company to make better decisions as to how quickly the instruments are to be delivered.

Generally, the faster the shipping method, the more expensive it will be. Ground transportation is less expensive than air freight. Also, the bigger the order, the lower the "per unit" shipping cost. It can take almost the same amount of paperwork to ship a pallet of instruments as it does to ship one instrument. On a large shipment the cost of preparing the shipping documents, the invoice, and the accounting is spread across many items instead of one. With a little forethought and, to be realistic, with a fair amount of luck, sending large orders early enough to allow ground transport should realize some savings in shipping costs.

When you are preparing a requisition or otherwise asking purchasing to get something delivered to your site, please be aware of the cost to deliver something quickly. If you don't need it tomorrow or next week, tell purchasing and they will arrange for ground shipment.

Special handling charges can be significant. Suppliers will understandably want to share the extra cost with the purchaser if they have to disturb their normal production runs to manufacture your device in a hurry, outside their normal production run. If you are able, leave plenty of time to purchase your control system devices.

The bid provided by your friendly neighborhood instrument supplier should indicate how freight will be handled. Some of the common freight payment terms are listed in Common Freight Payment Terms table.

Common Freight Payment Terms	
PPA	Prepay and Add – The supplier pays the shipping, but the cost is then added to your invoice so the purchaser (you) ultimately pays for freight. This is probably the most common freight payment method.
Allowed	Allowed – The supplier pays for shipping.
Collect	Collect – The shipping company will collect the costs upon delivery.
Own Truck	Own Truck – There are several variations on the term, but it means that the instrument supplier will deliver it to you. There is usually no cost, unless a cost is stated.
Will Call	Will Call – You will pick up the instrument at the supplier's location. Many suppliers have a door that is actually marked "Will Call". Will Call is, of course, more common for smaller components that can be hand carried, and when your local supplier is truly local.
Owner's Truck	Owner's Truck – again, there are many variations on this term, but it means you will pick up the device. Similar to Will Call, but used more for larger components that require a truck and loading dock.

Instrument suppliers can be understandably reluctant to use Freight Allowed terms, since they can't predict if expensive overnight shipping will be called for later—after they have committed to a price. Shipping Collect is more expensive than other pre-paid methods, in part because shipping companies need to be paid for the time and effort expended to collect money at the receiving end, and for the possibility that no one will pay for the shipping after the fact.

Prepay and Add (PPA) freight is common. The supplier pays freight costs at the time of shipment. This approach likely gets the best shipping rates. The buyer agrees ahead of time to pay the shipping costs, which are added to the invoice by the supplier. The supplier is happy because they don't have to deal with unpredictable shipping costs and the risk is low that they won't be paid; remember they did a credit check on your company before they submitted a bid.

Step 6: Award the purchase order.

The purchasing group will probably make the award based upon the results of the technical and commercial evaluations. Some observations about the process:

a. It is always a good idea for you to review the specifics of the purchase order against the Specification Forms and the bid evaluation. It is not unheard of for items to get forgotten, for substitutions to be made, or for other problems to arise. A cursory glance at the documents may allow you to catch problems early, while they are still small.

b. The delivery schedule, delivery date, ship date, or variations on those terms means the estimated time the supplier feels they need to produce the component or equipment covered by the bid. The time is measured from the day the supplier formally receives the order to, probably, the day the equipment ships from the factory. Use of the word "date" in the term for this duration during the bidding process is ill advised, in our opinion, because the committed duration depends on difficult to control variables such as the date the supplier "receives the order" – is a phone notification of the purchase order sufficient or does the supplier require a written and signed purchase order? If your procedures include the submittal of information from the equipment supplier for review and comment by the owner prior to entry of the order, this process will consume a significant amount of time, often more than expected. The purchasing staff can work with the supplier and the owner to set mutually acceptable and reliable dates and durations.

   You may notice the letters ARO associated with the shipping duration in a bid; the letters mean "After Receipt of Order". Sometimes that means "Written Order," as mentioned before so the lag in mail delivery may become important. There are times when a phone call by your pur-

Payment Terms

Control system engineers and maintenance personnel normally have little or no involvement with the legal contents of a purchase order. However, for the curious, the following explanation discusses the process and some of the common terms used.

The "terms and conditions" is part of a purchase order. Indeed the "T&Cs" are frequently pre-printed on the purchase order form itself. They establish a legal agreement between your company and the company that supplies the device. You, as the maintenance or design person, may be asked to comment upon the duration of the device warranty, or you may be involved with setting up the required delivery schedule. But other than those "component usage"-based aspects of the agreement between the two companies, there may be little need for control system personnel involvement.

However, if you are in a hurry for your instruments, you will want to encourage your purchasing group to ensure that the terms and conditions, or T&Cs, or any other legal aspects of the purchasing process, are in order. It is appropriate to take great care when the devices you want to purchase are critical to your schedule—or if you are new to the bidding and purchase of your devices.

Disagreements between the buyer and seller are not uncommon, and they can add weeks to the acquisition process. Preparing a formal "Request for Quotation" package that includes final Specification Forms and a copy of your company's standard terms and conditions is good way to reveal potential problem areas early in the bidding process, leaving time for the purchasing and legal departments to resolve any conflicts.

A word of warning when working with a design or construction contractor: don't assume that your company's purchasing relationship with the supplier will apply when the devices are purchased by another entity—even if it is for your facility. The company selling the devices is only concerned with whose name is on the purchase order, not with the destination. It may not matter if your company has a long history with that supplier; the legal agreement between the seller and the buyer, in this case the contractor, will still need to be worked out. Credit checks will be run, prior history will be reviewed. The discount pricing that your company enjoys will have to be negotiated if the devices are bought under someone else's purchase order.

Working with a construction contractor can be interesting. Sometimes the contractor may end up with better pricing for a specific project than your company receives, since sometimes they buy a lot more devices than your company does. The authors have been in the situation where the contractor had to work with the owner to have the owner clear an outstanding debt before the instrument supplier would ship the instruments.

The payment obligation term appears as some variation of the term "Net 30." Net 30 means the buyer will pay for the device within 30 days of invoice. "Invoice" is the bill for the device, which is probably sent to your company the same day the device is shipped. The "Terms and Conditions" may key off other dates—for instance, the payment duration may commence the day the device ships from the factory or it may start the day you receive the device, regardless of the day the invoice arrives. Some payment schedules, particularly for more expensive or fabricated items, may call for staged payments based on some fixed duration after receipt of the order (ARO). For example, the supplier might look for 15% of the order value in payment two weeks after receipt of the order (to cover design costs), 30% four weeks after order (possibly to cover raw material costs), etc.

chasing professional might be sufficient to start the order, but that implies a history or a relationship between the buyer and the seller. In many cases, having a purchase order number is sufficient for the seller to start working on the order. Whatever the circumstances, these opportunities for misunderstanding between the buyer and the seller as to the delivery schedule should all be worked out prior to the award of the order. Many times the project construction schedule is off by 3 or 4 weeks due to the time it takes for the purchase order to be received and the manufacturing and delivery clock to actually start.

c. The quoted shipping duration was probably based on the Ex-Works date, also called the "Ship date". Note that "Ship" or "Ex-works" date is not a "Received By" date. Yet, amazingly, people may use the "Ship Date" as the date the instruments will be available for use, or to be the same as the "Received By" date in the schedule. This is very common. There is frequently a one- to two-week built-in schedule bust, or error, when this happens. You can work with the people publishing your project schedule to add a delivery time estimate to the ship dates before publishing the schedule, or make sure the end users at your facility know to add a week or so to the ship date for their planning. When your instruments are traveling across the county, same-day shipping and receiving is pretty unlikely.

Step 7: Receive the devices.

Your facility or project probably has some sort of receiving procedure in place to keep track of delivered components. There are receiving reports and logs to be filled in and notifications to be made. There should be a written record that the shipment was inspected for compliance with the purchase order and for damage. Documents have to be sent to the Accounts Payable department.

If you are working on a project that is part of an existing facility, there can be conflicts between the project group and the receiving group; the project group might want all materials for their work received and documented separately at a segregated facility so they are readily available to the project team, while the receiving group may want all materials received through their normal channels and at their normal location so they are part of the in-place tracking system.

To prevent conflict, these groups should talk to each other! Detailed procedures and arrangements for the receiving and distribution of delivered components should be defined and agreed upon in advance of any material deliveries. There should be a list of purchase orders and a way to keep track of individual items on each order, not every order is complete with one shipment. Control system devices are easy to track, since they are uniquely identified with a tag number. The receiving process involves noting the purchase order and the tag number of all devices received, then getting that information to Accounts Payable. The accounting group will pay the invoice after the components are received in accordance with the purchase order terms and conditions. The control system group might be asked to verify specification compliance for the store-room person that receives the device since the specifications can be complex. The purchasing department will ensure their receiving logs are accurate so they are a good resource for delivery information. Someone will be responsible to ensure appropriate tracking information is maintained when, and by whom, instruments are removed from the storage area. The date a device is received is important since the invoicing process or the warranty might start upon delivery.

Instruments are frequently shipped with the operation manuals in the shipping container. Care should be taken to ensure that the manuals are delivered to the correct parties. In the authors' experience, copies of manuals should go primarily to the plant startup personnel and the owner maintenance personnel, followed with the delivery of any extra sets to the construction personnel. The construction personnel need Installation Details and Loop Diagrams for mounting the devices; operation manuals are of secondary importance. It is the plant startup personnel and the owner maintenance personnel who need the detailed information in the manuals. Understand, too, that the purchase order may have included a limited number of manuals; replacement manuals can be expensive, although the trend now is towards free access to manuals that are downloaded using the Internet.

Conclusion

In the previous pages, the seven steps of acquisition have been discussed at some length. Problems can easily occur if the control system group, the purchasing group, and the owner representative do not cooperate and communicate during the control system device selection and acquisition process. The authors know of instances where entire requisitions were never processed and where items somehow mysteriously disappeared from purchase orders. Perhaps the epitome of failure-to-communicate was a project where the entire set of control system devices was ordered, not once but twice. Of course, one set was surplus upon receipt at the job site. Unfortunately, there were also some "surplus" personnel after this occurrence. It didn't have to happen.

CHAPTER SIX

Binary Logic Systems

Overview

Defining and documenting binary logic systems so that all the stakeholders in the operation of a facility can understand and use the information is challenging. Each reviewer has a different skill set and knowledge of process control and yet each person has a need to glean the appropriate information to meet his or her job requirements. The documents used to define the control functions will have to be understandable to management, the process designer, the operations staff, the maintenance technician, and the electrical and control system professionals, the logic device programmers and the Supervisory Control and Data Acquisition (SCADA) system configurators. There are a number of documents that have been developed to meet this wide range of needs. This chapter addresses those documents.

As stated earlier, this book is based on ISA-5.1, the most recent edition of which now includes elements of both ISA-5.2-1976 (R1992) *Binary Logic Diagrams for Process Operations* and SAMA (Scientific Apparatus Makers Association) PMC 22.1 *Functional Diagramming of Instrument and Control Systems*. With this second edition, we have broadened the scope of this chapter to include more discussion regarding the contents of these documents.

There are a number of terms used when working with binary logic systems that are interchangeable. They include binary, discrete, on-off and probably others. Before we delve into this subject, here are a few definitions:

Discrete control is defined in the *ISA Dictionary* as on-off control. On-off control is defined in the *ISA Dictionary* as a simple form of control whereby the control variable is switched fully ON or fully OFF in response to the process variable meeting, or passing, a predetermined value.

The term "Binary" is defined in ISA-5.1 as a signal or device that has only two discrete positions or states and when used in its simplest form, as "binary signal" as opposed to "analog signal", the term denotes an "on-off" or "high-low" state.

Analog is defined in ISA-5.1 as a signal or device that has no discrete positions or states and changes value as its input changes value and when used in its simplest form as in "analog signal" as opposed to "binary signal", the term denotes a continuously varying quantity.

Most continuous process control schemes include some binary or on-off control. Batch processes are primarily this type of control. The control steps might result from the action of a switch or it may entail a long series of steps performed by a complex automatic system. On-off control can start and stop a motor or it can be used to initiate an orderly shutdown of an entire plant upon the detection of an unsafe condition.

In any facility, a simple binary logic control system might, for example, back up analog tank level control to ensure that a tank does not overflow. A complex binary logic control system might scan for many different process conditions and equipment states, and take action to protect major equipment worth perhaps billions of dollars from damage. Some binary logic systems are software-based and others are hardwired. Simple or complex, every system needs to be documented. Any proposed binary logic control scheme needs to be presented for review, discussion and implementation. The proposed controls interrelationships must be recorded so programming personnel can understand the system's intended operation. P&IDs were developed to show a continuous process, which they do very well. However, an augmented presentation method may be called for when showing binary logic systems.

ISA-5.1 includes several examples in Annex B of how combinations of analog on-off control schemes can be defined by the use of:

- Process control descriptions
- Instrument diagrams
- Functional diagrams
- Electrical schematic diagrams

Each documentation method may be used by itself, or, more frequently, they will be used together to meet disparate requirements at various stages during design and operation. There will be widely different levels of skill and understanding among the persons using binary logic documentation for review, implementation and training. This alone may drive the types of drawings and text used and the standards to which they are developed.

The success that each method has in showing how the logic works is also dependent on the complexity of the control needed. The ability of the designer to document the system concisely and accurately and the ability of the process owner, maintenance technician, and even the programmer to read the system documentation correctly contribute to understanding.

Process Control Descriptions

Complex logic requires more detailed descriptions than those provided on a P&ID. Furthermore, technicians doing the control computer programming and configuration normally need more detailed information than that provided with a few notes on a drawing. For this reason, a text-based description may also appear on a separate document called a Process Control Description, a Functional Specification or an Operation Description. There are as many names for these as there are companies that use and produce them. A text-based logic document is typically prepared in support of each process control system.

A typical process control description for an entire system will include several sections. ISA-5.1 shows this two-part process control description in paragraph B.13.1. The two-part approach is particularly effective as it provides different levels of detail to meet the needs of different reader ability and interest.

1. Process Description

   A. The system title

   B. A general description – A prose description of what the system does, including establishment of the system limits, that is, where in the system the document starts and where it ends. It describes the major components that make up the system and in basic terms describes how the equipment is intended to work. The document addresses the conversion of the material inputs to the product output. On design projects, the project scope can be a useful starting point for this portion of the Process Description.

   C. Form – The Process Description is an executive summary. It is intended to be read and understood by the largest number of people. As stated earlier in this chapter, this section of the document tends to be more conversational in nature. Specific ranges and set points for individual loops do not appear in this section, although the operating range and limits of the overall system are appropriate data.

2. Control Description

   A. Provides more specific data on how the control system will perform.

   B. Written as a loop-by-loop description, providing detail on each loop that makes up the system.

   C. Format – This section uses more stylized prose which can, or should, translate into specific logic steps or modules in the program. The detailed description typically requires some familiarity or experience with process control, and it probably will be challenging for the casual reader.

Additional information is sometimes included in a process control description including:

- Referenced P&IDs – The drawing numbers and titles.
- Motor List – A listing of motors and the interlocks that impact their operation.
- Loop List – A listing of analog and discrete loops affected by on-off control elements.

There are many different applications of the text-based descriptions of logic. The document should be written by an experienced control professional. Programmers will use the document as the basis of design. Statement brevity is good; remember that binary logic is on-off, so the statements should reflect that. Similar functions should be written the same way every time, hence the "stylized" writing statement above. The programmer will likely execute the logic the same way every time, so the English language description of that logic step should be written the same way every time.

The Control Description of 5.1, Appendix B, paragraph 13 follows. The additional detail needed for the Control Description is shown:

1) Control system design for:

a) Small volumes for long and short periods should allow tank to fill to a high level to automatically start the pump and then to stop the pump at a low level.

b) Large volumes for long periods should allow the pump to run continuously and maintain a fixed level with a level-to-flow cascade control loop.

2) Pump (run, or operation) control is selected by a three-position Hand-Off-Auto (H-O-A) selector switch:

a) Selector switch is in "HAND" position.

b) Selector switch is in "AUTO" position.

3) Pump should be stopped at any time:

a) Automatically if low level is exceeded.

b) By operating the stop pushbutton.

c) Switching the H-O-A selector to "OFF" position.

The example is very simple. In practice, control descriptions can be quite complex, but it is necessary to bridge the gap between the process designer's intent and the programmer's interpretation, and, importantly, to capture the details so someone can understand how the program was written long after the process designer and the programmer have moved on to new glories.

Instrument Diagrams

ISA-5.1 states that an instrument diagram, shown in Figure 6-1, may be used to define a binary logic system. The diagram describes the process using the symbols for the control devices and the process equipment described in Chapter 2. The instrument diagram contains much of the information on a typical P&ID. Instrument diagrams become more common for complex control applications or where the site standards only allow for minimal control information on P&IDs. There is implied, process control knowledge with their use.

Figure 6-1: Instrument Diagram

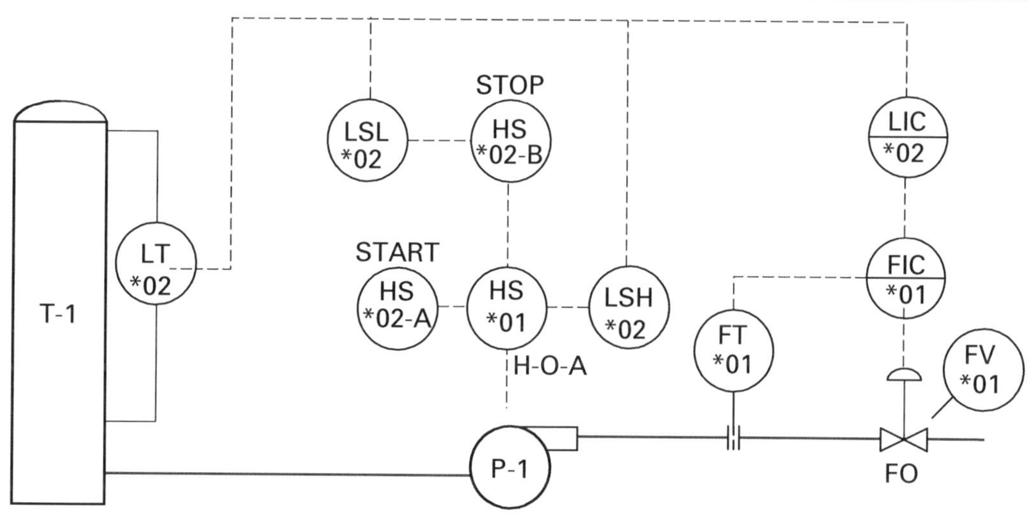

From ANSI/ISA-5.1-2009

Functional Diagrams

Another method from ISA-5.1 for describing a binary logic system is a functional diagram. The functional diagram included as Figure 6-2 repeats the control scheme shown in the instrument diagram but using different symbology. The symbols are those originally developed by SAMA as part of PMC 22.1 *Functional Diagramming of Instrument and Control Systems*. See Chapter 10 and Figure 6-3.

Readers are encouraged to study ANSI/ISA-5.06.01-2007 *Functional Requirements Documentation for Control Software Applications* which provides systematic development ideas and examples on this subject.

Figure 6-2: Functional Diagram

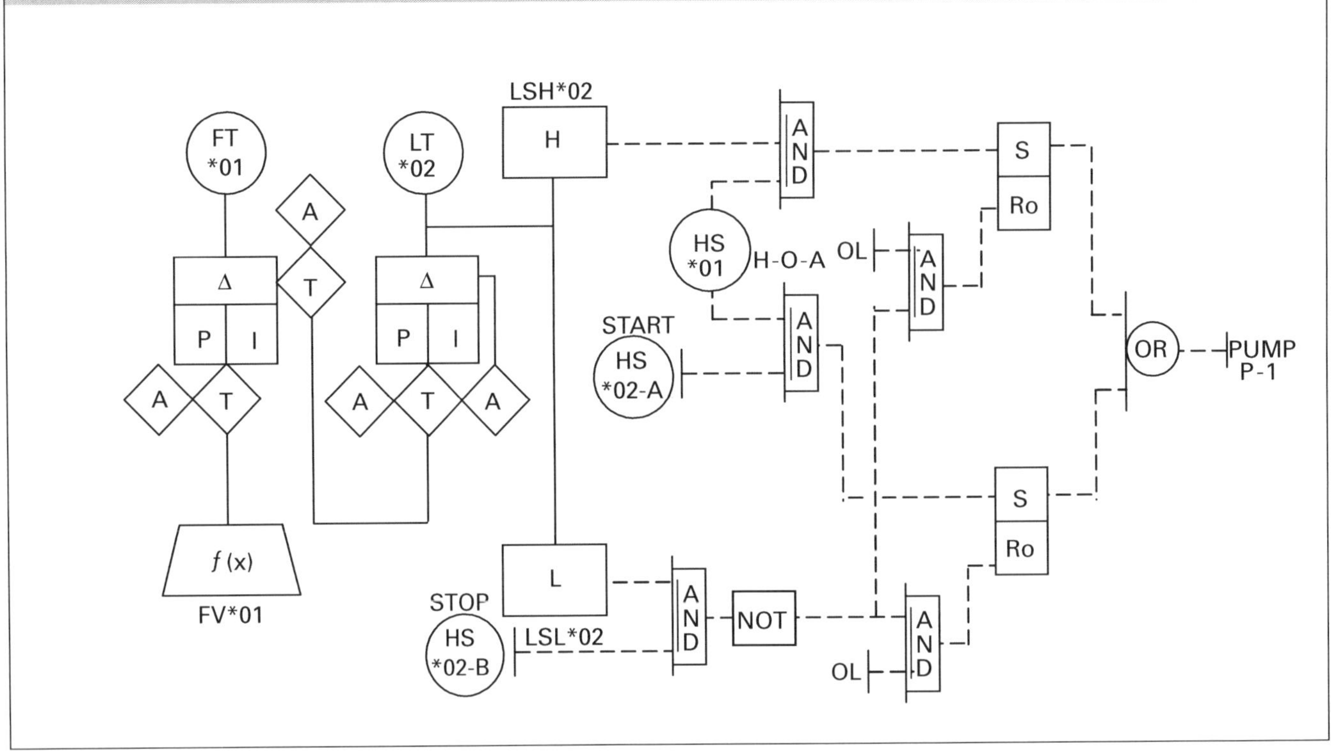

From ANSI/ISA-5.1-2009

Electrical Schematic Diagram

The electrical schematic diagram included as Figure 6-4 shows how the control scheme can be defined and wired. The diagram defines how the high level switch (LSH*02) and the low level switch (LSL*02), the start button (HS*02A), the stop button (HS*02B), the motor starter (M), the starter contacts (M1 and M2), the overload relays (OL), and the hand-off-automatic switch (H-O-A) can be wired to satisfy the process control description.

There are other methods of defining binary logic systems. Some of these are:

- Ladder diagrams
- Interlock notes
- Logic diagrams

Ladder Diagrams

A ladder diagram is defined in the *ISA Dictionary* as follows: A diagram used to program a programmable controller. They are stylized schematics representing power flowing through a network of relay contacts arranged in horizontal rows called "rungs", between two vertical rails on the side of the diagram that contains the symbolic power. The basic program elements are contacts and coils as in electromechanical logic systems.

Figure 6-3: Functional Diagraming Symbols

No	Symbol	Description
1	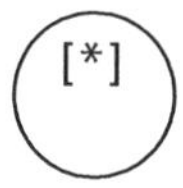	• Measuring, input, or readout device. • [*] = Instrument tag number. • Symbols from Table 5.2.1 may be used.
2	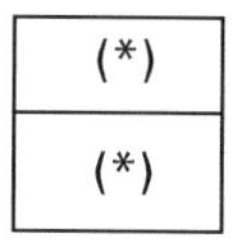	• Automatic single-mode controller.
3	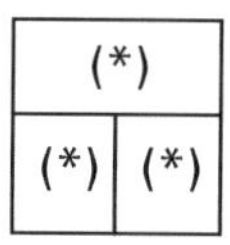	• Automatic two-mode controller.
4	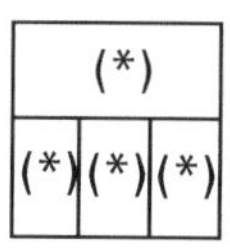	• Automatic three mode controller.
5	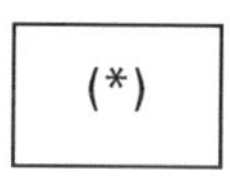	• Automatic signal processor.
6		• Manual signal processor.
7	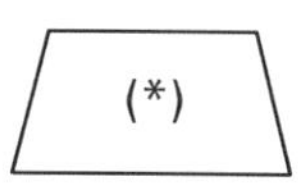	• Final control element. • Control valve.
8	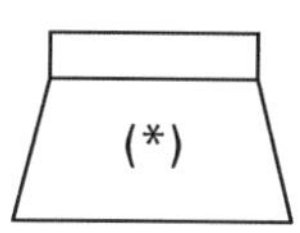	• Final control element with positioner. • Control valve with positioner.

From ANSI/ISA-5.1-2009

Figure 6-4: Electric Schematic Diagram

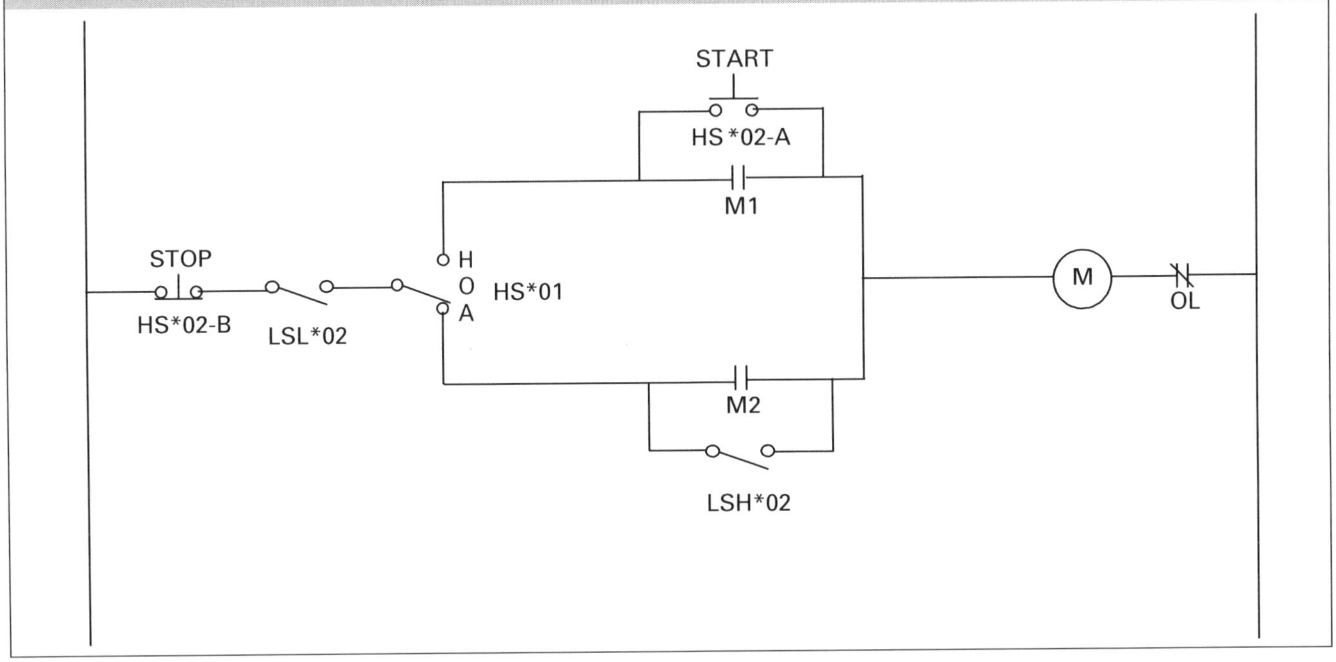

From ANSI/ISA-5.1-2009

Ladder diagrams can be used to program programmable logic controllers (PLCs) or the on-off control for some distributed control systems (DCSs), or to define the logic of hardwired relays. The vertical lines connected by horizontal logic circuit lines have a ladder-like appearance, hence the name. Learning to read and use ladder diagrams is a useful tool for understanding the design of logic circuits.

The traditional ladder diagram for programming discrete control is becoming less common with the advent of features packed programming tools for logic devices like PLCs. These new tools use visual elements that represent defined or established blocks of processing tasks (process blocks). These elements are interconnected using identifiers that may or may not include ISA tags, depending on the system used and the programmers.

A process block (program element) can have simple or quite complex features. The advantage of this approach is that a single program element or process block can represent many lines of a ladder diagram, and the settings that are entered by the programmer are frequently clearly shown. Once you understand the function of a particular element or block, it is easy to understand how it works for your process. You then copy the block to different applications as needed. The control platform manufacturer supplies standard blocks or elements for your use, and most also allow the user to develop their own. It is a powerful and quick way to program. The programmer essentially connects the dots between program elements. Of course understanding which points are to be connected and which program elements should be used is where the skill of the programmer becomes evident.

Documenting your operating logic system by printing out the PLC program is simple and accurate, although it can use up reams of paper. For ladder diagrams, there is the drawback that not everyone working at a process facility is able to read and understand them. If programming notes are not included with the ladder diagrams, it is likely that understanding a complex logic set will be frustrating. On a PLC ladder diagram, the individual action elements, called "coils," can be referenced by many individual rungs, so reading the printout can call for thumbing through hundreds of pages of printout. This can be confusing, even when the printouts are indexed properly.

In addition to the methods of defining on-off control described above, other methods are used, including Interlock Notes and Logic Diagrams. They are described below.

Interlock Notes

Interlock is defined in the *ISA Dictionary* (definition 2) as a physical device, equipment, or software routine that prevents an operation from beginning or changing function until some condition or set of conditions are fulfilled.

The interaction of the equipment and the action of final control elements in response to the sensed process conditions can be described with short statements that briefly describe the action and the reaction (response). This set of logic statements, often called "interlock notes," will typically appear on the P&ID along the right side of the drawing or in a section along the bottom. A symbol with a number is used to identify the separate statements. A numbered symbol, typically a diamond, may also appear on the P&ID next to each of the elements affected; for example, the process sensing switches, transmitters, valves and motors.

This is an effective method to use because of its clarity and the immediacy of the information. The method facilitates drawing reviews and, more importantly, it is a good system for jogging memories years later. The operating description is carried along with the P&ID in perpetuity. Drawbacks to the method are due to the necessary brevity of the statements; it is not appropriate to have detailed statements in interlock notes as the drawing real estate required is better used by other drawing elements. There is not a lot of space available on an already busy P&ID. The interlock notes need to be short. The detail provided is intended to provide a sense of what a specific interlock does; when more detail is desired the reviewer knows to find the logic diagram, which presents the complete story.

Logic Diagrams

The Logic Diagram is a versatile method of depicting on-off control. The symbols are simple enough to understand, yet very complex operations can be shown efficiently. All the on-off control elements for a single piece of equipment can be shown clearly on one drawing. Figure 6-5 contains a definition of a Logic Diagram.

Figure 6-5: Logic Diagram Definition

- The Logic Diagram is a conceptual document that defines the on-off state of a process, and depicts the scheme necessary for control.

Logic Diagrams use a series of symbols to indicate what is happening in an on-off system. While there are many variations of Logic Diagrams, there is much similarity among them. The basis for the information shown is ISA-5.1 and ISA-5.2-1976 (R1992), *Binary Logic Diagrams for Process Operations*.

Figure 6-6 shows two of the symbols used to describe functions in Logic Diagrams: AND and OR. A Logic Diagram is set up with the inputs or actions on the left side of the drawing and the result, or results, on the right side. The AND symbol signifies that all inputs must exist (or all actions must be taken) before the result occurs. A more formal definition that might be drawn from Figure 6-6 is that result "C" occurs if and only if "A" and "B" exist or if action "A" and action "B" have both been taken. If there are more actions feeding into an AND, all actions must have taken place to get the desired result. There is no limit to the number of actions feeding into an AND.

The OR symbol signifies that one or more inputs must exist (or one or more actions must be taken) if the result is to occur. A more formal definition that might be drawn from Figure 6-6 is that result "C" occurs if, and only if, "A" and/or "B" exist or if action "A" and/or "B" have been taken. If there are actions "A" through "Z" feeding into an OR, one or more of the actions must have taken place to get the desired result.

Figure 6-7 shows the NOT gate and the Basic memory symbols. The NOT symbol reverses the input. If the action has taken place and is fed through a NOT, no result will occur. If the action has not taken place and is fed through a NOT, the result will occur. A formal definition that might be drawn from Figure 6-7 is that "B" exists if and only if "A" does not exist.

The Basic memory symbol is more complex. If an action has been taken, the result will occur and continue to occur until another action takes place. The symbol has two outputs, and they flip-flop. If one shows the action, the other will show no result.

The formal definition from Figure 6-7 is that "C" exists as soon as "A" exists and continues independent of "A" until "B" exists. "D" exists when "C" does not.

We have shown only a few symbols, ISA-5.1 includes many more.

Figure 6-6: Binary Logic Symbols – AND & OR

No	Function / Symbol / Truth Table	Definition / Graph

1 — AND gate

- Output true only if all inputs are true.
- Alternate symbol

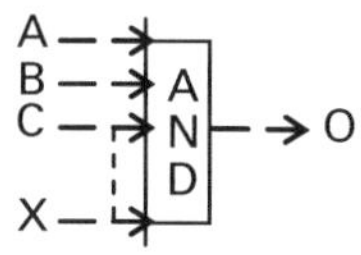

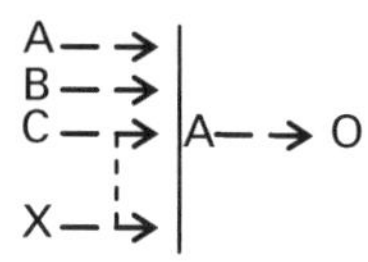

	A	B	C	X	O
1	0	0	0	0	0
2	1	0	0	0	0
3	0	1	0	0	0
4	0	0	1	0	0
5	0	0	0	1	0
6	1	1	0	0	0
7	1	0	1	0	0
8	1	0	0	1	0
9	0	1	1	0	0
10	0	1	0	1	0
11	0	0	1	1	0
12	1	1	1	0	0
13	1	1	0	1	0
14	1	0	1	1	0
15	0	1	1	1	0
16	1	1	1	1	1

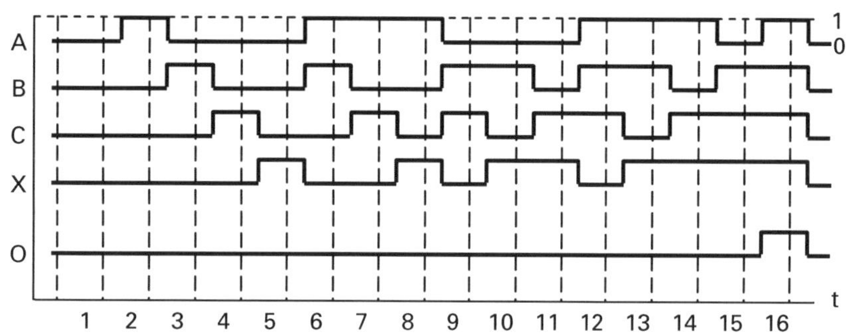

2 — OR gate

- Output true if any input is true.
- Alternate symbol

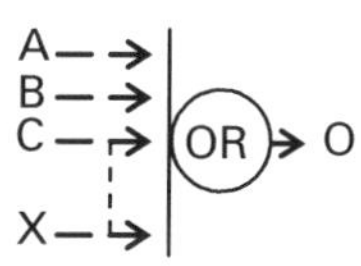

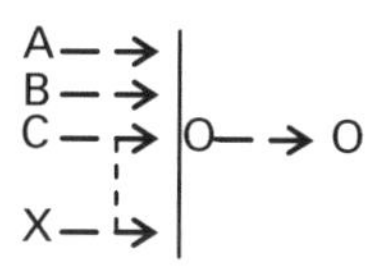

	A	B	C	X	O
1	0	0	0	0	0
2	1	0	0	0	1
3	0	1	0	0	1
4	0	0	1	0	1
5	0	0	0	1	1
6	1	1	0	0	1
7	1	0	1	0	1
8	1	0	0	1	1
9	0	1	1	0	1
10	0	1	0	1	1
11	0	0	1	1	1
12	1	1	1	0	1
13	1	1	0	1	1
14	1	0	1	1	1
15	0	1	1	1	1
16	1	1	1	1	1

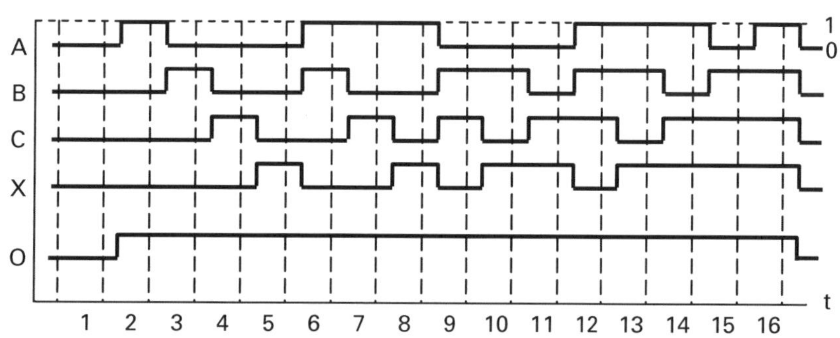

From ANSI/ISA-5.1-2009

Figure 6-7: Binary Logic Symbols – NOT & Basic Memory

No	Function / Symbol / Truth Table	Definition / Graph
11	NOT gate A --- [NOT] --- O	• Output false if input true. • Output true if input false.
	A | O 1 | 0 0 | 1	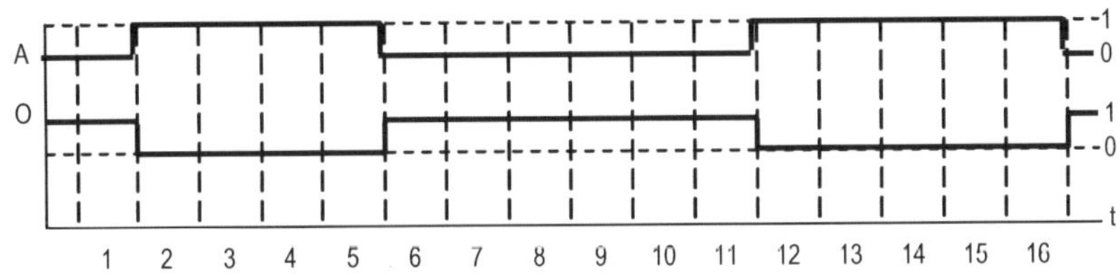
12	Basic memory A → [S] → C B → [R] → D	• Outputs [C] and [D] are always opposite. • If input [A] equals (1) then output [C] equals (1) and output [D] equals (0). • If input [A] changes to (0) output [C] remains (1) until input [B] equals (1) then output [C] equals (1) and output [D] equals (0). • If input [B] equals (1) then output [D] equals (1) and output [C] equals (0). • If input [B] changes to (0) output [D] remains (1) until input [A] equals (1), then output [D] equals (1) and output [C] equals (0). • If inputs [A] and [B] are simultaneously equal to (1) then outputs [C] and [D] change state
	(truth table below)	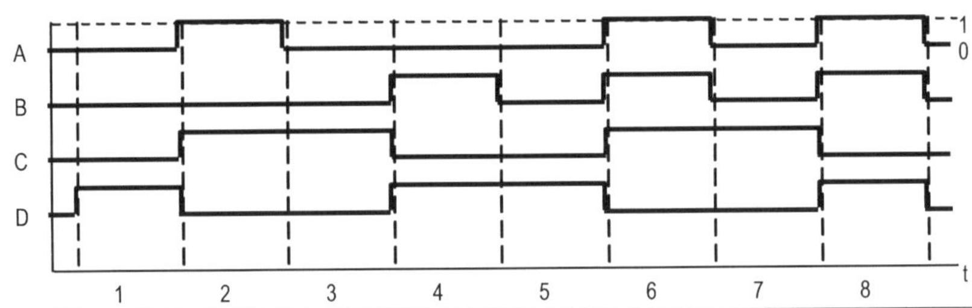

	A	B	C	D
1	0	0	0	1
2	1	0	1	0
3	0	0	1	0
4	0	1	0	1
5	0	0	0	1
6	1	1	1	0
7	0	0	1	0
8	1	1	0	1

From ANSI/ISA-5.1-2009

Binary Logic Implementation

Several devices that can perform on-off control have been mentioned, including relays, PLCs and even DCSs. The oldest form of electronic on-off control, relay logic, was accomplished by wiring electro-mechanical relays into control circuits. Relays were great for the purpose but they take up a fair bit of room, contacts and coils can fail, they generate a lot of heat, and they can make complex on-off control schemes difficult to implement and revise.

With the advent of electronic data processing came computers to perform the same functions as relays. The most common and versatile of these is the programmable logic controller, or PLC. These ubiquitous systems have software-based relays and coils that implement the Logic Diagrams through programming. They are relatively inexpensive and reliable.

As the cost of computational power decreases, more DCSs and bus-based control systems (digital communication based control systems), are implementing on-off control as well. Some of these systems use ladder diagram programming, while others use function-block-based programming. Suffice it to say that function blocks are programmed using a set of symbols that, once understood, can be read and interpreted relatively easily. Electronic controllers such as single loop digital controllers (SLDC) have some integral on-off control capabilities. Some field transmitters even have on-off control capability. The on-off control capabilities of electronic configurable devices are almost limitless.

Documenting On-Off Control

OSHA specifies that safety-oriented systems be properly documented, including any shutdown and interlock systems that are designed to mitigate personnel and equipment losses.

During control system design, the logic documentation is a coordination tool between the different design disciplines and system users. The equipment group, the process engineers, the operators, and maybe the chemical engineers all have input into how the system is to be operated. For the purposes of this discussion, let us call them all "process engineers."

Ideally, their input should be recorded in a manner that is easy for them to use and understand. Frequently, the process engineers work with the control systems design group to establish the entire control system—both the analog and the on-off control—using the P&IDs and one or more of the forms of logic documentation. Later, the programmers work with the control systems design group to implement the logic, sometimes long after the process engineers have defined the intended operation. When the electrical group handles on-off control, the Logic Diagram can be the interface enabling the electrical group and the control systems group to complete the design of the interlock system cooperatively.

An Example Logic Diagram

Figure 6-8 illustrates how a Logic Diagram may be used to define an on-off control system.

This figure includes a common motor start circuit, as it would appear on an electrical drawing. To start the motor, someone depresses the start button located next to the motor or on the starter housing itself. If the overload relay contacts are closed—that is, if the motor is not overloaded—this action completes the circuit to the coil in the motor contactor. As the coil is energized, coil contact "C" closes, or "seals" the start, and the motor starts. To stop the

Figure 6-8: Motor Start Circuit and Logic

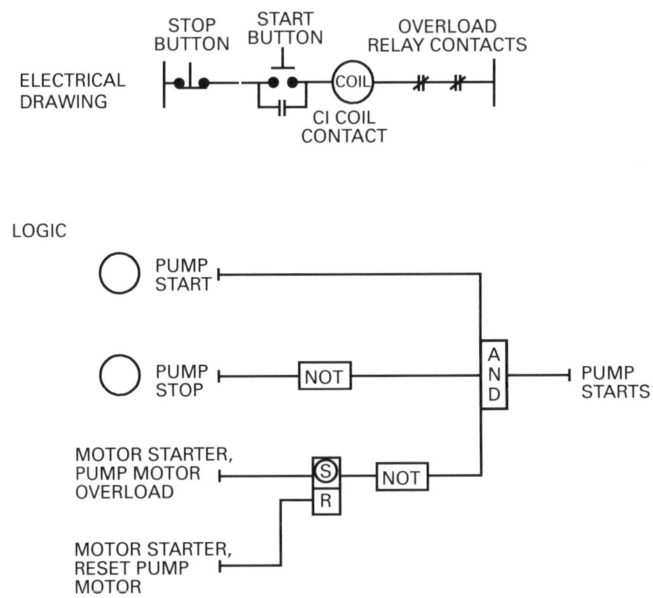

motor, the stop button is depressed. This interrupts the circuit to the contactor coil, which opens the contacts and removes power to the motor. Figure 6-8 also shows how this same action is shown on a Logic Diagram.

For the pump to start, all three inputs to the AND are necessary. The first action is to actuate the start button. The next action (as "action" is used in connection with binary logic) is not to actuate the stop button. This action is reversed by the NOT, and so the top two inputs are satisfied. Including the "not stopped" allows you to have a separate action to stop the motor. The motor is not overloaded; there is no input, and therefore, no output. However, there is a NOT in the line which reverses the "no output" and the pump starts.

In the MEMORY, a small o is placed around or next to the S. This means that the starter overload relay overrides the starter reset. In real terms, this means that if the starter overload relay shows an overload, the reset button will not start or jog the motor.

Developing a Logic Diagram

Next, we will develop a Logic Diagram for our example project. Please refer to Figure 2-21 in Chapter 2, P&IDs. There are three diamonds with an internal 1 on the P&ID. One is on the output of LSL-301, the low-level switch on 01-D-001. The second is on the start-stop scheme (HS/HL-401 and 402) of 01-G-005, the condensate pump. The third is on HV-400, the hand operated on-off valve. The diamond is on the output of ZSH-400, the position switch high which signifies that HV-400 is fully open.

The details are shown in Figure 6-9 Logic Diagram L-1. L-1 has five inputs to the AND. All five must be satisfied before the pump will operate.

Figure 6-9: Logic Diagram L-1

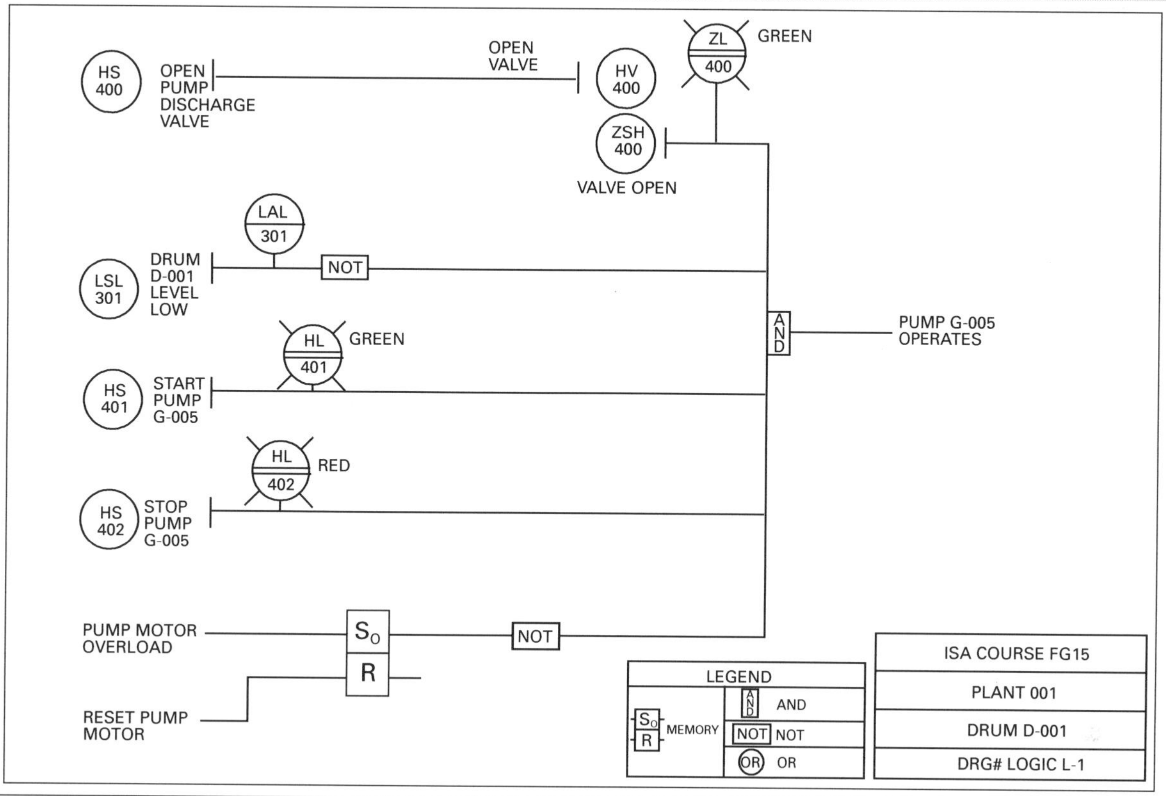

Starting at the top, we depress HS-400 to open HV-400. There is a short time lag before the valve is proved open by ZSH-400 and ZL-400 is lit.

The next input LSL-301 and LAL-301 are not activated, that is, there is liquid in the tank. This action is reversed by the NOT and the second input is satisfied.

Next, pump start switch HS-401 is depressed, green light HL-401 is lit, and this input is satisfied.

The fourth input HS-402, the pump stop switch, is not depressed. This action is reversed by the NOT and the fourth input is satisfied.

The last input to MEMORY is not activated, so there is no output from MEMORY. This action is reversed by the NOT, so this input is satisfied and the pump starts.

Logic Diagram L-1 shows how a simple on-off system works. For a simple system, a word description might tell the same story: In order for pump G-005 to start, the start button must be depressed, there must be liquid in D-001, the stop button must not be depressed, and the pump motor must not be overloaded. However, when the system is complex, a word description can easily lead to misunderstandings and possibly unsafe conditions. Therefore, in the opinion of many control system specialists - but certainly not all - the use of Logic Diagrams in connection with on-off will remove ambiguities and therefore make the plant safer. The use of Logic Diagrams does, however, require training a critical mass of people so the knowledge of how to understand the documents is spread throughout the necesary groups in your organization. Members of the design teams and the owner's engineering and operations personnel must be trained to understand and use Logic Diagrams.

There are two types of on-off control systems. First, there is on-off control to run the normal operation of the plant. There are motors to start and stop and equipment to run. Failure of these control schemes is certainly undesirable. A failed on-off control scheme might result in a troublesome spill or produce off-specification product. These results are annoying and expensive, but no one is hurt or at risk. No equipment is destroyed. To illustrate, we will assume that we can implement this type of on-off control by use of a basic process control system (BPCS).

The second category introduces a different responsibility for on-off control systems—those that are safety related. The documentation may appear quite the same, but importance of these systems is much greater. Unfortunately, the distinction between normal operating on-off control systems and safety systems can be fuzzy. One approach might be to use the regulatory documents to define which systems are truly safety systems and which are normal operating systems. There are many codes and standards that address on-off control. For instance, the National Fire Protection Association (NFPA) has published standards for the operation of many different kinds of burners. The standards describe in detail how the on-off control is to work. To illustrate, we will assume that we can implement a similar but not identical process using a safety instrumented system (SIS).

The *ISA Dictionary* defines an SIS as: A system that is composed of sensors, logic solvers, and final control elements whose purpose is to take the process to a safe state when predetermined conditions are violated...other terms commonly used include emergency shutdown system (ESS), safety shutdown system (SSD), and safety interlock system."

From the P&ID we can see there is some sort of SIS involving a low level in 01-D-001, the condensate pump 01-G-005, and the on-off valve HV-400.

For improved safety, it is often desirable to separate the basic control scheme from the SIS. ANSI/ISA-84.00.01-2004 *Functional Safety: Safety Instrumented Systems for the Process Industry Sector* states in annex B.1.1: "Separation between BPCS (Basic Process Control System) and SIS functions reduces the probability that both control and safety functions become unavailable at the same time, or that inadvertent changes affect the safety functionality of the SIS. Therefore it is generally necessary to provide separation between the BPCS and SIS functions."

Some processes are well known, and what was included in the last plant design is often the basis for the current project. The previous experience of some team members might suggest that certain parts of the process need special care. Books and technical papers have been written, describing certain processes and how they can be operated safely. Our project team has decided that reactor control will be safer if the BPCS is separated from the SIS, so we have developed two slightly different control systems to control a reactor.

Let's start by showing how a project might be implemented as an integral part of the BPCS (Figure 6-10).

Figure 6-10: Integrated Control System

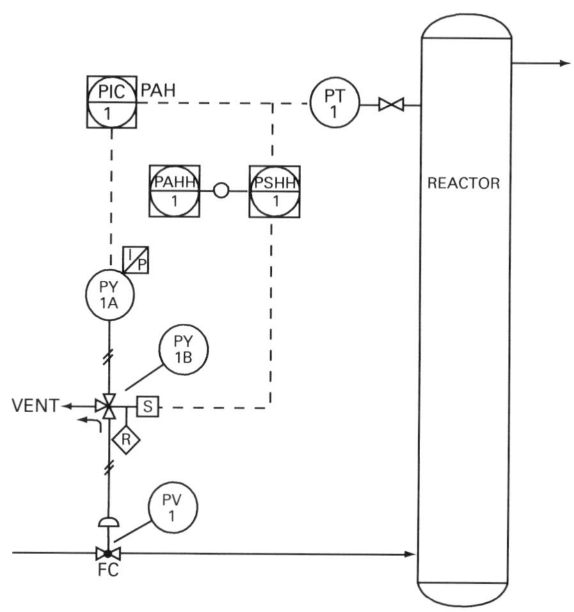

This figure shows a simple control scheme that employs the capabilities of a shared display-shared control (SD/SC) system or a DCS or a programmable electronic system (PES) as the basic process control system (BPCS).

The SD/SC shuts off the feed to the reactor when the reactor pressure gets too high. When the system is operating normally, PIC-1 modulates the position of control valve PV-1, which keeps the reactor pressure at the set point. If the reactor pressure rises above the set point of a high-pressure alarm (PAH-1 – part of the BPCS), PAH-1 activates. If the reactor pressure continues to rise, a high-high pressure alarm, PAHH-1, activates. Concurrently, the power to solenoid valve PY-1B is interrupted and PY-1B moves to its fail position. PY-1B shuts off the signal to PV-1 and opens the ports to permit the pneumatic pressure on PV-1 to vent to atmosphere. Since PV-1 is a fail closed valve (FC), it closes shutting off the feed to the reactor. PY-1B stays in the failed position until the pressure in the reactor decreases below the alarm point of pressure switch PSHH-1 and until the solenoid is reset locally and manually, as indicated by the "R" within the diamond.

It is apparent this system will control the reactor pressure as long as the BPCS is operational. However, a separated system will make the overall system safer.

Figure 6-11 shows how the system might be defined using Safety Instrumented System symbology. A new pressure switch, PSHH-2, is added to the scheme. This switch is connected directly to the reactor. In addition, solenoid PY-2 now controls the pneumatic pressure to a new control valve, PV-2. This control valve is fitted with a position switch ZSL-3 to confirm that the valve is closed. In normal operation, loop PIC-1 controls the pressure in the reactor. An alarm sounds if the reactor pressure rises above the set point of PAH-1, which is an integral part of the BPCS. If the reactor pressure continues to rise to the alarm point of PSH-2, valve PV-2 closes, the feed stops and the reactor pressure will rise no longer. When the reactor pressure is again normal, solenoid PY-2 can be manually reset and the control system is operational.

Figure 6-11: Separated SIS System

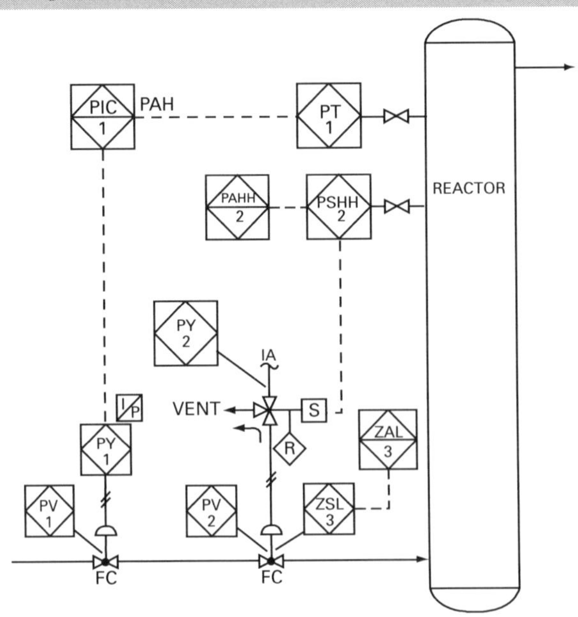

In this chapter we have used simple written descriptions to define the intended operation of a reactor SIS and BPCS system. There are, of course, other ways to present this information. ANSI/ISA-5.06.1-2007, *Functional Requirements Documentation for Control Software Applications*, addresses in more detail the methods and documents that can be used to accomplish this task. The standard also covers the validation and testing of control software.

At least two methods are used to determine how dangerous a process is. The two are described in ANSI/ISA-84.00.01-2004 Part 3 (IEC 61511-3 Mod). The standard uses the terms semi-quantitative and semi-qualitative, which in discussions of SIS are often shortened to simply "quantitative" and "qualitative."

Semi-quantitative: For example, a specified consequence should not occur with a frequency greater than once in a hundred years, or some other frequency.

The quantitative method develops mathematical answers to the following questions:

- What is the risk of failure, that is, what is the probability of a system failing to respond to a demand?
- What are the consequences of that failure? If the system does not respond correctly, what happens?
- Are these consequences more serious than is acceptable? If yes, then develop a new SIS to reduce the risk, and repeat the process. If no, proceed with the implementation of the SIS as designed.

Semi-qualitative: What is the risk? As "risk" is used in this specific context: the risk is a combination of the probability and severity of harm and is dependent on the consequences of the hazard, the probability that the hazardous area is occupied, the probability of avoiding the hazard, and the number of times per year that the hazard would occur if a safety function was not present.

The ANSI/ISA standard helps in reaching numerical answers to the above questions. The quantitative method may sound difficult to implement. However, some firms that have adopted this method are reporting excellent results. Some firms also are reporting lower costs for safety systems and safer plants. The emphasis is placed where the consequences are the most severe by the use of a mathematical solution to the analysis.

Summary

In this chapter we have looked at on-off control in a continuous process plant and how the Logic Diagram can help in our understanding of how a SIS (Safety Instrumented System) works. We have described what on-off control is and how process control descriptions, instrument diagrams, functional diagrams and electrical schematic diagrams can document it. We have studied examples of Logic Diagrams and pictured some of the symbols that may appear on one.

The electrical drawing has been compared with the Logic Diagram for a motor start circuit. An integrated BPCS for a reactor has been described, and how the same reactor might be controlled by a SIS.

We have also briefly discussed the quantitative and the qualitative methods of developing an SIS using ANSI/ISA-84.00.01-2004 Part 3 (IEC 61511-3 Mod) as a basis.

CHAPTER SEVEN

Loop Diagrams

Overview

The Loop Diagram is surely the most recognizable document used in the control system field. It is the most common one as well, except perhaps for the Specification Form (instrument data sheet). The popularity of a well-executed Loop Diagram stems from its universal utility. Loop Diagrams are found on designers' desks, in construction trailers, in notebooks in the maintenance shop or stuffed into the pockets of engineers and maintenance technicians, and can even be found littering the bottoms of field junction boxes. They are spread throughout industrial facilities because they are useful! Loop Diagrams tell us the components that comprise our control system and how they are interconnected, in a stylized but logical and somewhat elegant fashion. Keep in mind, however, that the presentation of information on Loop Diagrams, and, indeed the information itself, is undergoing some change and development as the systems to which they apply change.

The *ISA Dictionary* definition of a Loop Diagram is:

> "A schematic representation of a complete hydraulic, electric, magnetic or pneumatic circuit." The complete circuit is generally called a loop.

Definitions of a loop from ISA-5.1, paragraph 3.1.40 are:

> "Instrumentation arranged as a combination of two or more instruments or functions arranged so the signals pass from one to another for the purpose of measurement and indication or control of a process variable."

> "A self-contained device that measures and controls a process variable."

These definitions are broad and cover any interconnection between two or more devices (except, of course, for the special case of a single self-contained device) for the purpose of measurement and indication or control of a process variable.

The key concept in the definition of "loop" is that all the devices in the loop monitor and indicate or control a single process variable. Why? Well, because it makes sense, but that is not really a helpful answer. There are an awful lot of control system devices in most process control facilities. Having numerous items tagged with unique, random identifiers is probably sufficient to buy and install them, but at some point people are going to have to understand at a glance what the instruments do and whether or not they are related, and in what way: If I pull this fuse or disconnect this wire, what is going to happen and to what?

Having a drawing show that relationship and not having to wade through a pile of sequential connection diagrams is a real time saver. If you have ever worked with the connection diagrams prepared by, for example, a packaged equipment supplier (think of a self-contained compressor, boiler, or mixer) that does not use loops, you probably understand. Such documents are set up to facilitate quick assembly of the equipment, including options that you may not have, sufficient to wire to an I/O (input/output) card; they are not always set up to make it easy to understand how a component works in conjunction with another component in the equipment's control system. Now, add 20 other such islands of packaged equipment, each with a manufacturer with its own understanding of how to define and connect the devices to their control computer, and the familiar "Figure out what is going wrong and get it fixed, NOW" experience could become unpleasant. The loop concept captures the relationship between devices in easily understood groups, and the format and approach are fairly universally understood by control system personnel.

A simple loop might be just a field mounted pressure switch that illuminates a light on a local panel when the pressure in a pipe gets too high or, for you purists, when the process variable has exceeded the set point by some predefined amount. In this example, the loop only has one field instrument. More commonly there are more components: some combination of measurement, action, and process manipulation. A pipeline was chosen for this example because the authors wanted to show that proximity does not enter into the definition of a loop. A pipeline, such as one that conveys hydrocarbon liquids across the country, can cover long distances, miles even. It is common for the final control element that manipulates the process pressure in a pipeline to be in another operating unit, remote from the pressure measurement. Remember, the commonality that defines a loop is the single process variable, in this case the pressure in the pipe.

A common electronic control loop, particularly one performing continuous process control, might consist of a transmitter, a controller, an I/P (current-to-pressure) transducer, and a control valve. A more complex electronic system that is still a single loop might include a transmitter with two or more control valves with I/P transducers, as long as the valves exist to influence a single process variable. For example, in the pulp and paper industry, pulp consistency control can be set up this way; the pulp stock pumped from a chest or tank includes a consistency transmitter that sends the process variable to the distributed control system. A valve is manipulated by the control system to provide gross adjustments of water flow to meet the consistency set point; a smaller valve is also manipulated to adjust the consistency with more precision. All this hardware comprises one loop.

Figures 7-1 and 7-2 show typical loops.

Figure 7-1: A Pneumatic Loop

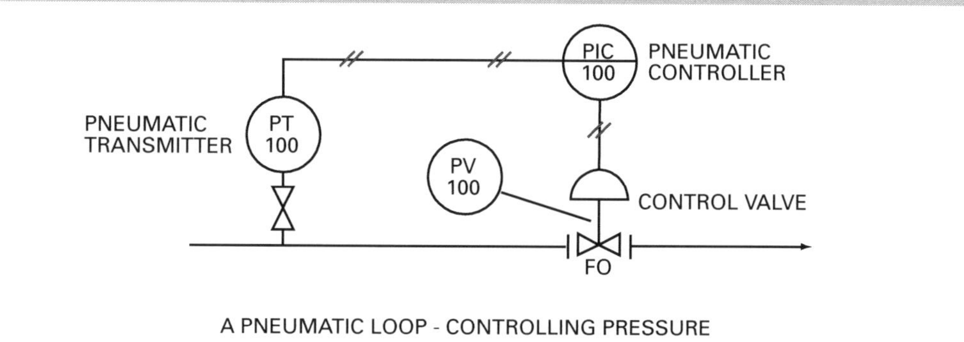

Figure 7-1 shows a pneumatic loop (P-100) controlling pressure in a pipeline. All signal transmission is pneumatic, most commonly at 3–15 psig.

Figure 7-2: An Electronic Loop

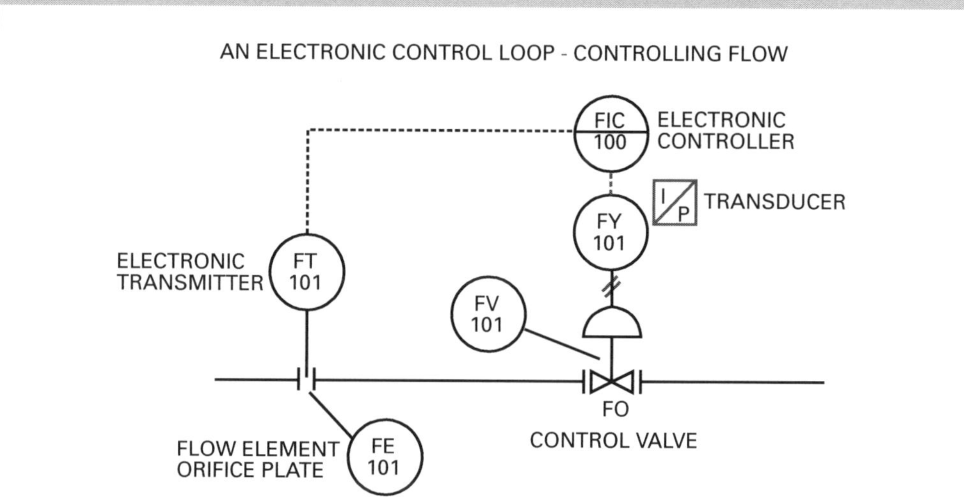

Figure 7-2 is an electronic flow loop (F-101). The transmitter and controller outputs are electronic signals, most commonly 4–20 mA. The transducer (FY 101) changes the electronic signal to pneumatic, typically 3–15 psig. Sometimes a higher-pressure signal, 6–30 psig, is used to provide more power to stroke (move) the control valve.

The loops are defined on the P&ID, meaning that the components that comprise a single loop are assigned during the development of the P&ID. Therefore, by the time the Loop Diagrams are developed, the components have already been identified with tag numbers.

Loop Sheet Size Variations – a Rant and Rave

You may be one of the unfortunates that have to live with Loop Diagrams published on a bed-sheet-sized drawing, ASME D or E: as in "ASME D or E" size, with 4 or 6 loops to a drawing. This defeats the primary benefit of the Loop Diagram – one loop per drawing. This is actually helpful to NO ONE, except possibly some long departed project design team. They had to follow an unfortunate contractual anomaly that priced drawings by size – they got paid less for six 11" x 17" drawings than they did for a 36" x 42" drawing.

Someone may have subjected your facility to this abomination because they thought it would save design dollars. What this really means is your design contractor had too much power or your management had too little understanding of process control requirements. So, you need to educate your management.

Luckily, with today's computer-aided drafting, the bed sheets can be cut up into individual Loop Diagrams fairly easily, but someone is going to have to pay for it!

So what comprises a Loop Diagram? A quick answer might be everything a control system technician needs to be able to install and maintain the instrumentation contained therein. All electrical and pneumatic signal and power connections for the measurement, indication and control devices should be shown and identified along with the connections to the process—to the piping, ductwork, vessel, etc. A Loop Diagram schematically defines the interconnection of each device needed to make the circuit work as intended, including the control computer or pneumatic controller. It identifies the source and contributions of service utilities like compressed air, electric power and water. A Loop Diagram provides termination information including junction box identifiers, terminal strip numbers, terminal block numbers and possibly even pneumatic port identification. Depending on the requirements of your facility, cable numbering, conductor colors, wire markers, power panel callouts and circuit breaker numbers all can be found on a Loop Diagram.

Guidelines for Loop Diagrams

ISA-5.4-1991 *Instrument Loop Diagrams* establishes a few guidelines for Loop Diagrams.

- Loop Diagrams are typically ASME B size (11" x 17") drawings. The B size drawing is used primarily because it has just the right amount of space to depict a single loop. This is a perfect size because it is easily handled, both in the field and on the technician's bench. The B size drawing can easily be reduced in a photocopier to ASME A size (8 1/2" x 11") for desktop reference and even storage in a three-ring binder. Printers that can produce 11" x 17" drawings are everywhere, so it probably will be simple to print off a copy of the Loop Diagram before you dash out to figure out why the operators lost the flow signal (probably because someone closed the upstream manual block valve or drove a forklift into a signal cable, if your day is anything like the authors').
- Loop Diagrams will depict only one loop; measurement of a single process variable, a control action, and manipulation of the process, or some combination of the three. Sometimes other drawings such as alarm panels or multipoint indicators might use an 11" x 17" drawing format; despite the loop-like appearance they are not since all the devices shown don't monitor or control *a single process variable.*

- Loop Diagram format may be either horizontal or vertical, but the chosen format should remain consistent for the facility or operating unit. Horizontal formats are more common; vertical formats can be read like a book when reduced to 8-1/2" x 11" size and bound in a notebook.
- Loop Diagrams should be divided into sections reflecting the component locations. Sections used in the Loop Diagram examples of ISA-5.4 include:

1. Field or Field Process Area	5. Panel Rear
2. Cable Spreading Room	6. Control Panel
3. Computer I/O Cabinet/Cabinet	7. Console
4. Panel Front	

The multi-section approach is one whose clarity and usefulness has been proven by long experience. The specific sections used in your facility may be different from those listed, but at the least a split between the field and the control panel is intuitive and useful in quickly understanding how the loop works. One section for field devices, an adjacent section or sections for Panel, or Panel Front (located in or on the front of the central or main panel or console) and Panel Rear (located at the rear of the central or main panel) is common. A third section showing the software functions; controllers or other actions has become common with the advent of distributed control systems (DCSs). Local (field or auxiliary location) control panels are frequently included in the field section rather than in a dedicated area. The goal is to pack all relevant information into a small area. Remember, relevancy is defined by the user: include information of use to the people maintaining the instruments.

The use of too many section breaks can cause inefficient use of space. Layout standards should be established with care. Three major sections is probably a decent compromise.

Your site standards for Loop Diagrams may call for loops to present a minimal amount of information, or they may be extremely complex and information packed. Minimum and optional information sets are discussed later in this chapter. The content of the Loop Diagrams at your facility must follow your site standard. The amount of information maintained on Loop Diagrams has a real, direct and significant impact on your facility's speed of response by maintenance staff to system problems. The facility technicians need a document that they can grab and go with, where the drawing provides enough information to get the problem fixed and the process back into control. Too much useless information will make the drawing difficult to read and will impact how quickly the drawing is understood. Plus, the more information the drawing carries, the higher the cost of design, so there are contractual and work-hour consequences to choices made regarding Loop Diagram complexity.

The loop drawing standards for your facility should establish the level of detail desired in depicting the software functions on a loop diagram. One approach is for the simple presentation of the software actions which assumes the reader will understand underlying functions, similar to that shown on some P&IDs. For example the loop sheet will just show the instrument bubble inside a square and the tag for an indicating controller. Not stated is the likelihood there are other functions that occur along with the basic controller such as summers, timers, rate of change functions, etc. For the higher level or more complete picture there is an implied need for another document to fully define the programming of the control device. Another approach for a loop diagram might be to show the minutia needed to completely define every aspect of the controller.

For a simple depiction, the functions shown tell maintenance staff what is being done with the signals without explaining how. It is likely that level of detail is sufficient for maintenance needs and even for construction. Even with the simple approach, when the signal is used by another loop, it is beneficial to add the reference loops so maintenance knows if they take down that loop; it may have an impact on other signals. The reference should be in the software area of the loop diagram because that is the level in which the relationship is made. The interaction can be easily defined with an arrow and the referenced loop number in text.

Detail can be added beyond the simple approach if desired, however it is recommended to consider who will make the most use of the loop diagram and if the information added is important to them. If there is another document such as a control schematic that shows the control information more completely, perhaps the complexities should be left off the loop diagram.

Here are some additional guidelines that are not directly addressed in ISA-5.4, but which the authors have found useful:

- A schematic representation of the process equipment to which the instrumentation is attached is helpful in both locating the devices and in quickly understanding the loop's function. This representation is part of the field section of the loop diagram. There is, of course, some design effort to produce this schematic and therefore there is a cost associated in showing the "process cartoon." The P&ID is a good source for the picture; however the designer should remove extraneous information to simplify the drawing.
- The title of the Loop Diagram is normally the service description that appears in the Instrument List, on Specification Forms and elsewhere on design documents and, sometimes, on the shared display screen. Do yourself a favor and make the title the same everywhere, there is no reason to "improve" from document to document. Using different titles for the same loop is very confusing to the people who use the documents.

- Stating what should be obvious, the identifying drawing number of a Loop Diagram should be the loop number, since that is the most natural way people will search for the document. Unfortunately, there can be resistance to this numbering scheme, since the loop number sometimes does not fit into the drawing numbering rules used for a project or a facility. In the authors' view, based on long experience, it is simply the responsibility of the control system design team to find a way to keep the Loop Diagram numbering system based on the loop number, as no other system is as useful for managing, locating and maintaining the drawing.

Optional Information

Sometimes the optional information specified in ISA-5.4 is included on the Loop Diagram. This can be particularly helpful for installing and troubleshooting, but again, there is a cost for adding and maintaining this information. Furthermore, adding this information to a Loop Diagram means that information appears in more than one place, the Loop Diagram and other project documents (perhaps the Specification Form). Duplicating information is normally a bad idea since corrections to the information become tedious and time confusing, or worse, the information ends up as different in as many places as it is used. Having information in two places, or more, means that at some point you will likely find the information to be correct nowhere; people neglect to correct the data everywhere. However, the advantage of having the manufacturer's data and range information on the Loop Diagram cannot be denied. Some facilities decide to maintain the Loop Diagram as the primary document for the control system group, so the Specification Forms become a secondary document used only for purchasing during project work. In other words, the Loop Diagram is the controlled document, and care is taken to ensure that its information is always correct. The Specification Forms may not be as carefully maintained (but they certainly should be).

Despite what you may have heard, device location information may be shown on a Loop Diagram, although this is not always done, nor is it called for in ISA-5.4. Hey, these are your drawings; you can put whatever you want on them. While it is not common, there are facilities that find floor and column location callouts for field devices to be useful. The information may be available from a Location Plan (see Chapter 8).

While this data can be helpful for the maintenance people at 4 a.m. when they are trying to track down a failed instrument, you must understand that location information is particularly expensive to provide and maintain. Device locations are not regularly available at the appropriate time in the design sequence. They may have to be added after the Loop Diagram is otherwise complete. The extra, out-of-sequence effort is expensive. Also, instruments tend to move during design and construction, so there is a surprisingly large cost to monitor and correct locations when they are provided on Loop Diagrams. This is probably better information for the Location Plan and database, but that is a decision to be made by the project team or facility management, with the understanding that there are contractual implications (which are code words for "expensive").

Additional information can be included on the Loop Diagram regarding the electronic or shared display (DCS, PLC, SCADA) when it is drawn using the ISA-5.4 definition of optional items. Therefore, it is important to know whether the Loop Diagrams in your design package include the minimum required information or the minimum plus ISA-5.4 optional items.

Loop Diagram information originates from the Instrument List, Specification Forms, Location Plan, Installation Details, and other project documents. All this information, and more, can and should be included in a project instrument database.

Loop Diagram Development

Now that you have an idea of what you want to include in the Loop Diagrams, it is time to make them. In general, Loop Diagram development can commence when:

- Design criteria for Loop Diagram content are agreed upon—in other words, you have a standard!
- P&IDs are issued for design.
- Specification Forms are complete.
- Devices have been specified or, better yet, purchased.
- Field junction box and marshalling panel architecture is understood and defined. Junction boxes and marshalling panels are interchangeable terms for the field panels where individual signal cables are gathered into one or a few large multiconductor cables for the home run to the host control panel; be it a PLC or DCS. These panels are not always used, but when they are, their presence should be included on a loop diagram since the degradation or disturbance of the terminations might be the reason for a loss of signal.
- Control computer termination diagrams are ready for "loading." Being ready for loading means you know what input and output cards will be used and you are ready to start populating the cards, or connecting devices to the points.

Typical Loop Diagrams

The following figures and discussions explain some typical Loop Diagrams and the information they contain. It should be noted that ISA-5.4 is the only generally accepted international standard that describes Loop Diagrams. It is not a complex document. It presents six typical Loop Diagrams and a few symbols used in their development. The six typical Loop Diagrams are of varying complexity and include two each for pneumatic control, electronic control and shared display and control. One of each type shows the minimum required items, and the other shows additional optional items. The symbols used on Loop Diagrams combine those used on P&IDs from ISA-5.1, but with additional electrical and pneumatic connection information.

The additional information is shown on the next three figures. Remember, when developing Loop Diagrams, the goal is to schematically depict ALL the connection points from the field devices to the process control device, be they pneumatic or electronic. All junction boxes, marshalling panels, pneumatic cabinets and terminations should be uniquely defined. All the connection information needed to install, maintain and troubleshoot the loop components should be shown.

Figure 7-3: Loop Diagram – Terminal Symbols

General terminal or bulkhead symbol

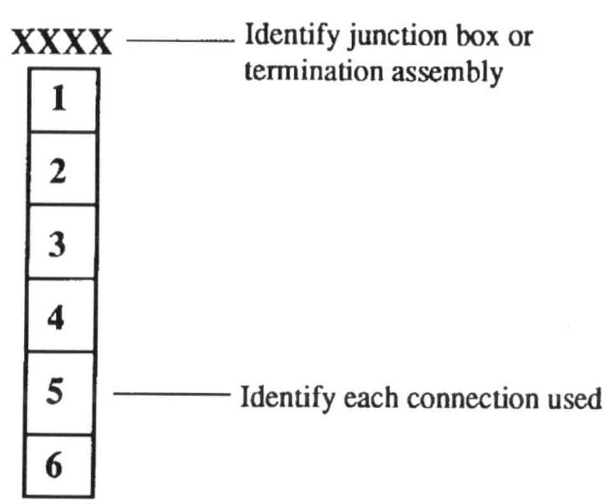

Instrument terminals or ports

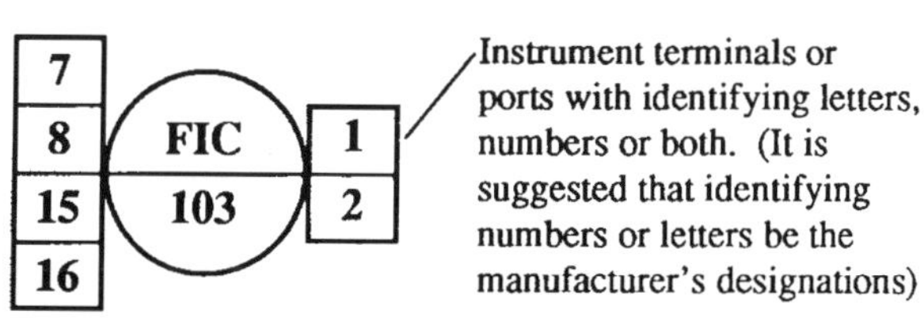

From ISA-5.4

Figure 7-3, Loop Diagram – Terminal Symbols, shows symbols used for terminals, bulkheads and ports. This should duplicate the identification placed on the device by the manufacturer. It is important that the termination callouts used on the Loop Diagram match those provided on the instrument. Showing + and – on the Loop Diagram may be correct for the 4–20 mA signal, but it won't be helpful when the device terminals are actually called out as 4 and 5, or when there is a "signal" + and – as well as a "test" + and –. The possibility of connecting to the wrong terminals may appear to be silly now, but it will happen nonetheless. Ambiguity will initiate misunderstanding. Misunderstanding during installation induces errors. Errors during installation cause the needless expenditure of VERY expensive hours, operations personnel goodwill, and maintenance technician energy during startup.

Figure 7-4: Loop Diagram – Energy Supply

Instrument system energy supply

Electrical power supply. Identify electrical power supply followed by the appropriate supply level identification and circuit number or disconnect identification.

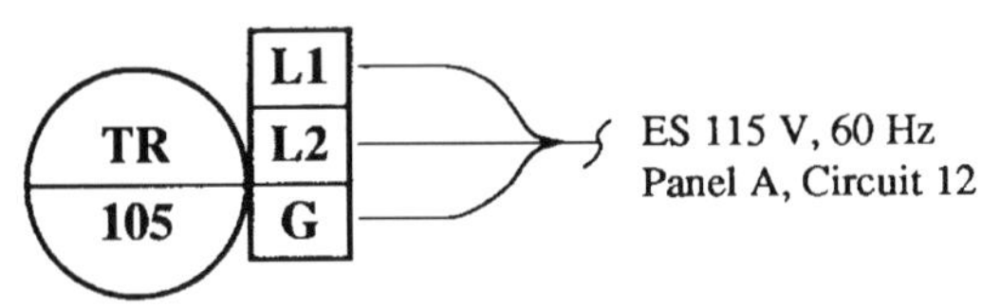

Air supply. Identify air supply followed by air supply pressure.

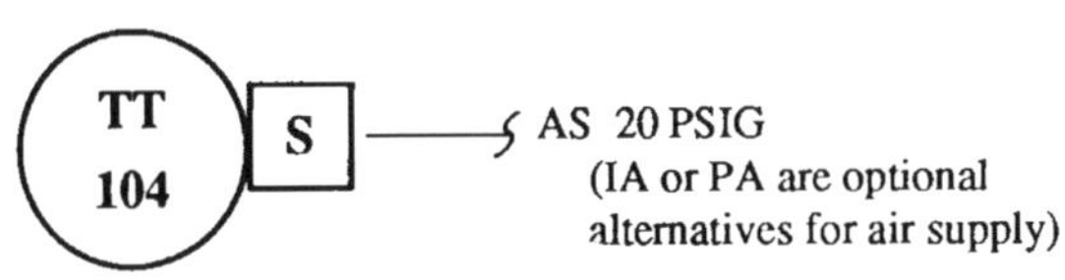

Hydraulic fluid supply. Identify hydraulic fluid followed by the fluid supply pressure.

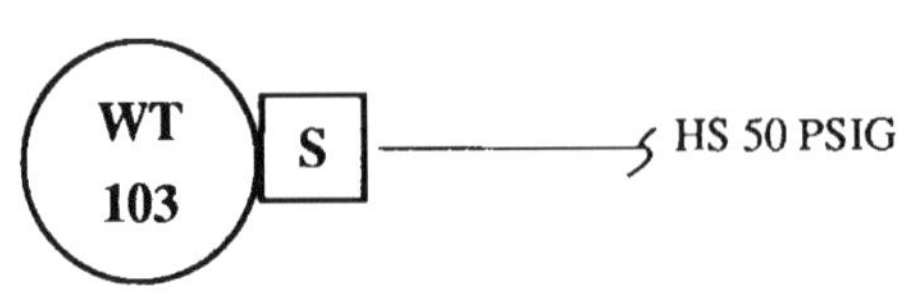

From ISA-5.4

Figure 7-4, Loop Diagram – Energy Supply shows how to connect the energy sources to the loop: the electrical source, air supply, and hydraulic fluid supply. Electrical information includes the number of conductors, voltage level, and panel and circuit numbers of the source. Air or hydraulic supply shows the number of connections and the supply pressure. Since the Loop Diagram is schematic, the physical mounting of the instruments is NOT shown. Physical requirements of the installation appear on another document called an Installation Detail (Chapter 8) which defines the physical layout and tubing size and fittings, wiring components and all other materials needed to mount and connect the devices.

Figure 7-5, Loop Diagram – Instrument Action identifies the response of the device to applied signals, be they pneumatic or electronic. A process transmitter or controller can be supplied in either of two actions, direct or reverse. In a direct-acting controller or transmitter, the output signal increases as the process variable increases. In a reverse-acting controller or transmitter, the output signal decreases as the process variable increases. The action for controllers should be indicated on a Loop Diagram, perhaps as DIR or REV. An incorrect controller action will make a control loop unstable. That instability may not be noticed until the loop is in operation. If the supplied action is incorrect, it is not difficult to reverse a controller action. The identification of instrument action for transmitters is not normally shown since they are almost all direct acting.

Figure 7-5: Loop Diagram – Instrument Action

Identification of instrument action. Show the direction of the instrument signal by placing appropriate letters close to the instrument bubble. Identify an instrument in which the value of the output signal increases or changes to its maximum value, as input (measured variable) increases by the letters "DIR". Identify an instrument in which the value of the output signal decreases or changes to its minimum value, as the value of the input (measured variable) increases by the letters "REV". However, since most transmitters are direct-acting, the designation DIR is optional for them.

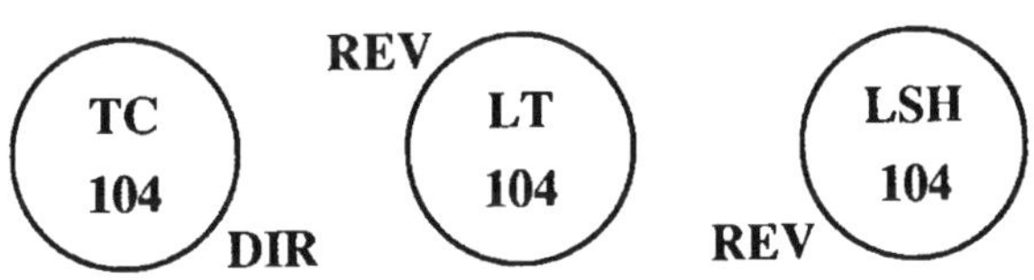

From ISA-5.4

Referring back to Figure 7-1, Pneumatic Loop P-100 will be used to illustrate the use of a direct or reverse-acting controller. Transmitter PT-100 is specified on its Specification Form as a direct-acting device. The output signal increases as the pressure in the line increases. Control valve PV-100 fails open; an increasing output from PIC-100 closes the valve. Therefore, the controller must have reverse action, and the loop will work as follows: As the pressure in the line rises, the output of PT-100 rises. Acting on this rise, the controller output decreases, opening control valve PV-100 and lowering pressure in the line.

Kids – Don't try this at home:

Many times, rules are appropriate. One situation that calls for a rule of thumb, or a rule in general, is the specification of valve positioner and I/P transducer action. In a modern control system, there are plenty of opportunities to fool with device action, at the transmitter, at the controller, at the valve. It is probably a really good idea to specify that all positioners and I/P transducers are direct acting, if only to limit the degrees of freedom in your system. You will appreciate this at 4 a.m. when trying to troubleshoot a misbehaving control loop. There is nothing worse than trying to get your head around a system with several changes in controller action from the transmitter to the control system, through the control algorithms, to the controller output, to the I/P transducer, to the valve positioner and then onto the valve itself.

Loop Diagram for P-100

For our project, we have developed a Loop Diagram – P-100, Figure 7-6. Figure 7-6 duplicates how loop P-100 might look on a P&ID. The DIR or REV is, therefore, not shown on Figure 7-6 and would not be shown on a P&ID.

Refer to the example P&ID presented back in Chapter 2, Figure 2-21. Included on that drawing is PIC-100, a pneumatic field mounted pressure controller that maintains the pressure in vessel 01-D-001 by releasing excess

Figure 7-6: Loop Diagram – P-100

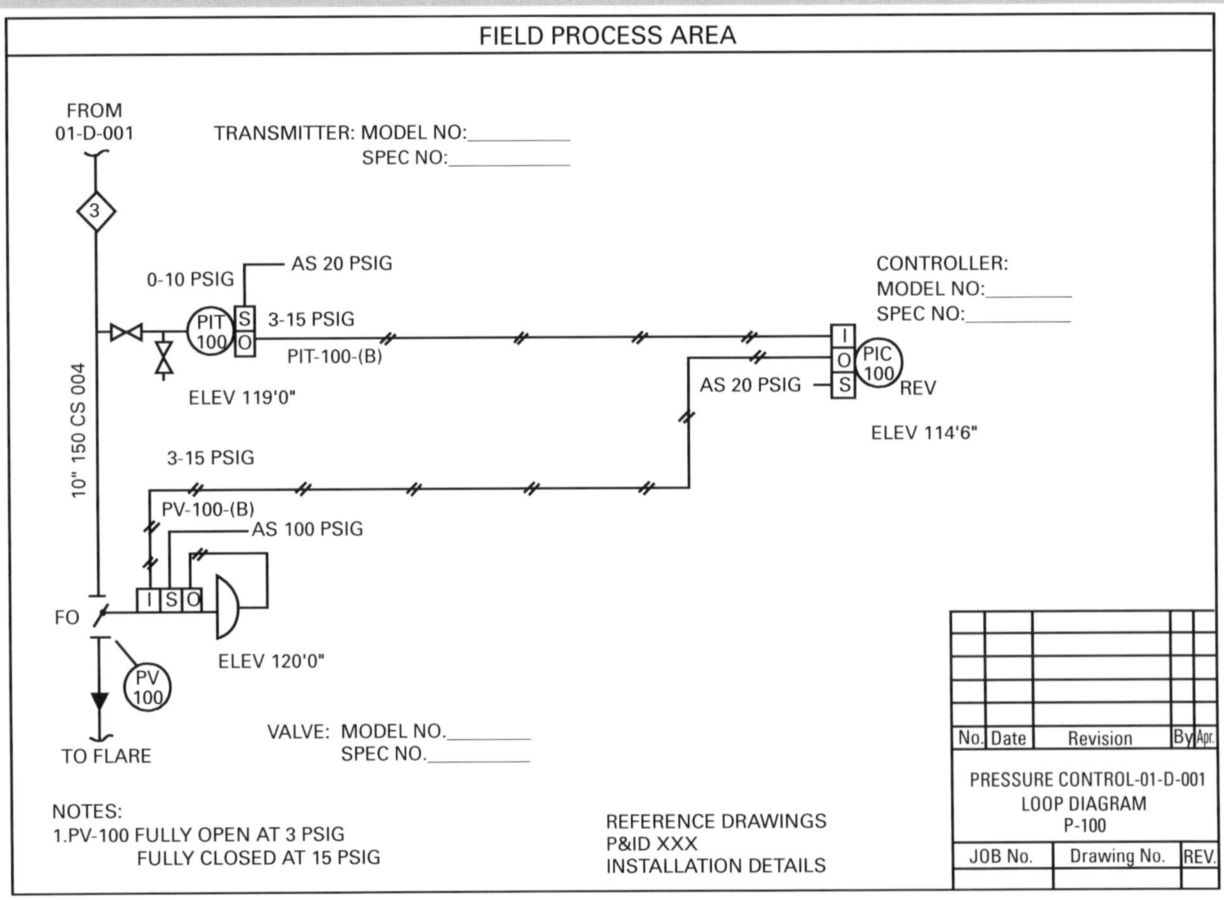

From ISA-5.4

vapors to the flare through butterfly control valve PV-100. From a Specification Form that we did not include, we nevertheless can expect to find the transmitter range since that is one of the fields that is to be filled out on an ISA form. This is discussed in more detail in Chapter 4. In our case the range will be 0–10 psig. Since ISA specification forms include line numbers, we can also cross check with the P&ID that PIT-100 is indeed on line 10" 150 CS 004. From the transmitter vendor data, also not included here, but you probably have seen a few, we find the transmitter has two ports; an air supply port S and a pneumatic output port O. The controller PIC-100 is located at elevation 114' 6"– N-1075' 0", E-1500' 0". How do we know this? The location and elevation as well as information for PIT-100 and PV-100 can be obtained from the Location Plan, see Chapter 8, Figure 8-4.

From the Specification Form we learn that PIC-100 is a reverse-acting controller with three ports. The vendor print shows the three ports, an input port I, output port O, and supply port S. The pneumatic output signal goes through tube number PV-100-(B) directly to the valve positioner port I. There are two other ports on the positioner, an output O and a supply S.

Note that a valve positioner does not show on the P&ID. It is only shown on the Loop Diagram. Valve positioners often do not have a unique tag number because they are mounted on, and supplied with, the associated control valves. The three air supplies originate at the six-connection manifold located at N 1060' 0", E 1500' 0" as shown on the Location Plan, Chapter 8, Figure 8-4.

The loop shown in Figure 7-6 works as follows: PIT-100, a direct-acting transmitter, sends an increasing signal as the pressure in 01-D-001 rises. The rising signal causes reverse acting controller PIC-100 to decrease its output to the direct acting positioner on butterfly control valve PV-100. Because PV-100 fails open, the decreasing signal opens the valve and lowers the pressure in 01-D-001.

To be honest, coordinate locations in text are not commonly provided, on Loop Diagrams or any other drawing. However, some firms are able to use the powerful tools of computer-aided drafting and some clever drawing manipulation to provide the locations essentially for free. The word "free" in this case is a bit disingenuous; someone had to write that clever software and its development costs were most assuredly not free. However once it is done it can provide the data for next to no cost. It is a neat feature if available, but the authors don't think you should get overly agitated if the option is not offered.

Figure 7-7: Loop Diagram – Electronic Minimum

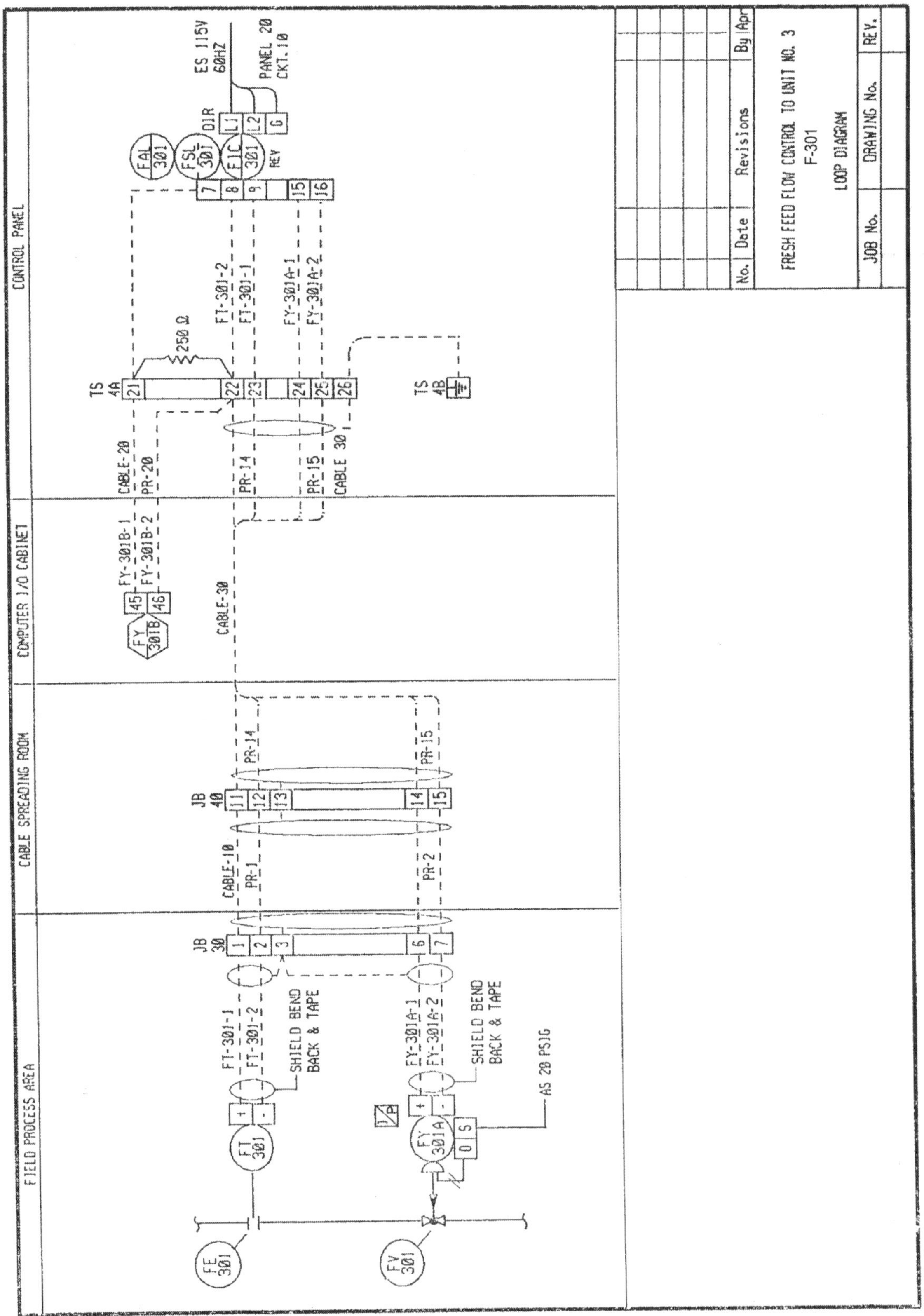

Loop diagram, electronic control, minimum required items

From ISA-5.4

Loop Diagram for F-301

Figure 7-7, Loop Diagram – Electronic Minimum shows an electronic loop F-301. On the left side of the drawing are the items in the field process area. The same tag numbers and symbols are used as are on the P&ID, but the symbols include additional details. Orifice plate FE-301, flow transmitter FT-301, and control valve FV-301 are shown. A pair of wires, FT-301-1 and FT-301-2, connect the + and – terminals of the transmitter to terminals 1 and 2 of field junction box JB-30. In our example we have verified that the instrument supplier really does mark the terminals with a + and –; it happens. There is a shield to protect the pair from electromagnetic interference. The shield is isolated at the transmitter as indicated by "shield bend back and tape" and it's connected to a common terminal 3 at JB-30.

A multiconductor cable 10 connects JB-30 with junction box JB-40 in the cable spreading room. The purpose of the cable spreading room is to reorganize the wires from field multiconductor cable 10 to an optimal arrangement for control panel multiconductor cable 30. The overall shields for the multiconductor cables are connected to terminal 13 of JB-40 to provide a continuous shield from the field to the control panel. The shield of cable 30 is connected to ground through terminal 26 on TS-4A, and from there to the ground terminal on TS-4B.

There is an input signal from FT-301 to computer FY-301B. The 250 ohm resistor connected to TS-4A terminals 21 and 22 converts the 4–20 mA signal of the flow transmitter to the 1–5 v DC signal needed by the computer. This signal uses pair 20 of cable 20 and wire numbers FY-301B-1 and -2. FIC-301 has a "behind the panel" integral low flow switch, FSL-301, and accompanying "front of panel" alarm FAL-301. FSL-301 is a direct-acting switch. As the flow signal decreases, the output decreases and the audible alarm does its thing. FIC-301 is powered by 115 v AC from circuit 10 of panel 20. The power supply is connected to FIC-301 terminals L1, L2 and G.

Controller FIC-301 is reverse-acting. A decrease in signal from FT-301 will result in an increasing output from the controller. The output of FIC-301 from terminals 15 and 16 of the controller is wired through TS-4A, terminals 24 and 25 to cable 30, pair 15. Pair 15 is connected to junction box JB-40, terminals 14 and 15. There it is connected to cable 10 pair 2 to junction box JB-30, terminals 6 and 7. From JB -30 a single pair goes to the I/P, FY-301A. The wire numbers of this pair are connected: FY-301A-1 to the plus terminal and wire FY-301A-2 to the negative. FY-301A requires a 20 psig air supply connected to supply port S, and the output connects under the diaphragm of the actuator. Control valve FV-301 will fail closed as indicated by the arrow pointing toward the valve body.

To show some differences in the details that can be provided on Loop Diagrams, Figure 7-8 is the same loop with some additional optional items included. Just to be different, wiring is depicted using the alternate (but not included in ISA-5.1) triple cross hatch rather than the more common dotted line used in Figure 7-7.

Figure 7-8: Loop Diagram – Electronic Minimum & Optional

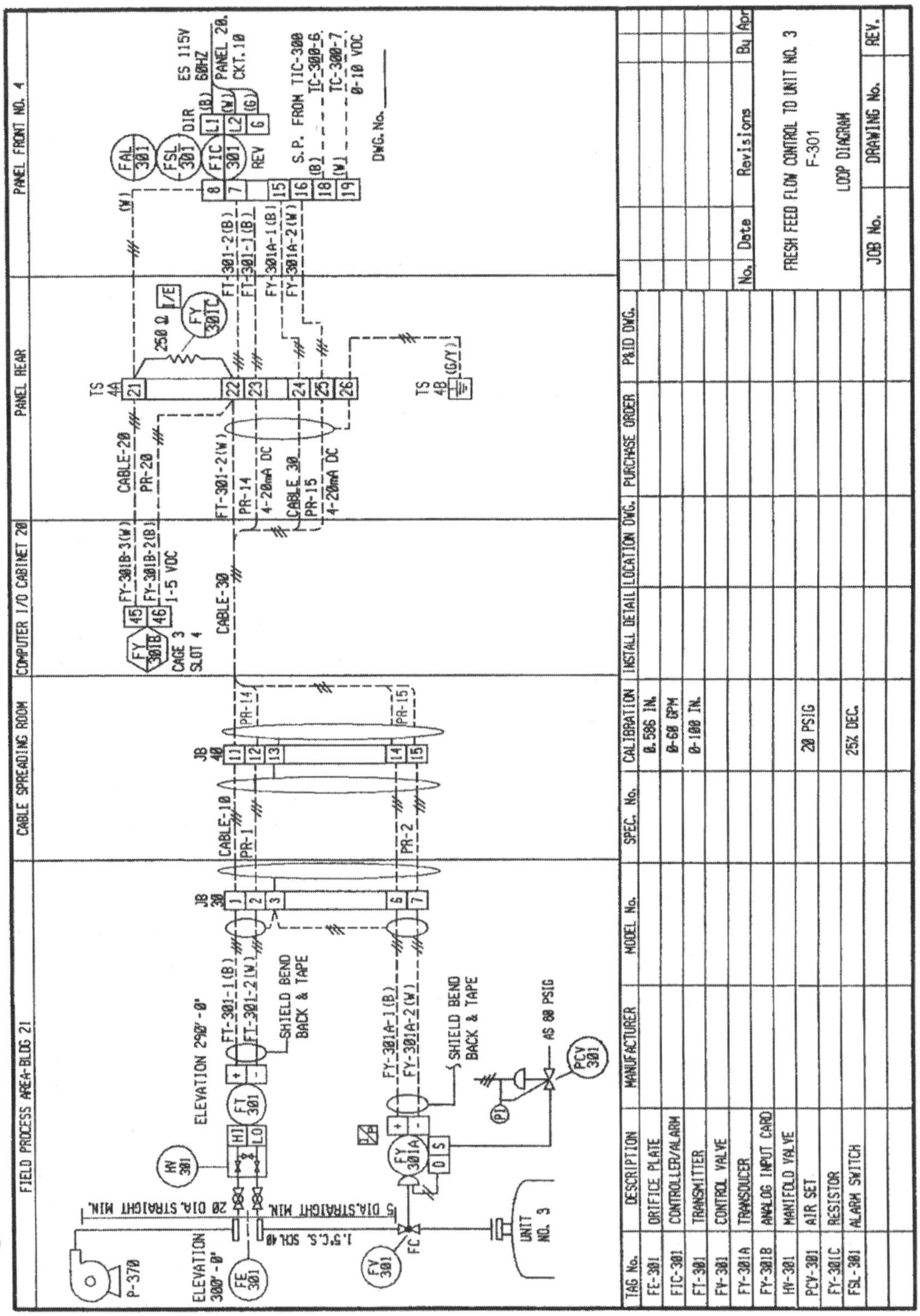

Loop diagram, electronic control, minimum required items plus optional items.

From ISA-5.4

The optional items shown on Figure 7-8, Loop Diagram – Electronic Minimum & Optional, include information about the pipe, 1.5" CS Sch 40, and other information:

- Its source is at pump 370 and its termination is in Unit 3.
- There are 20 diameters of straight run piping upstream of the orifice FE-301 and five diameters straight run downstream.
- The orifice plate is at elevation 300' 0".
- FT-301 is at elevation 290' 0".
- There is a three-valve manifold, HV-301, which permits the transmitter to be "zeroed" locally by shutting the upstream valve and opening the bypass valve.
- The voltage input of 1–5 v DC and the location information, Cage 3, Slot 4, are provided for the process control computer.
- Additional tagging has been added for three valve manifold HV-301, the 250 ohm resistor FY-301C, and the pressure regulator for the instrument air supply, PCV-301.
- The color of the panel wiring has been specified: "W" for white, "B" for black, and "G/Y" for green with yellow trace.
- Signal levels for the controller input and output are stated as 4–20 mA dc and the computer input as 1–5 v DC. Calling out the signal level may seem a bit obvious to someone who has worked with the same analog signal for the past 30 years, but there are many designers, installers and technicians who are dealing with mixed voltage, milliamp and fieldbus (digital) signals within their facilities. Calling out the signals may not seem so unnecessary to maintenance personnel who were just hired.
- One helpful feature is the table at the bottom of the Loop Diagram that lists the tag numbers and device specific information on all the instruments shown on the drawing.

Additional Feature		
Description	Spec number	Location Drawing
Manufacturer	Calibration	Purchase Order
Model Number	Installation Detail	P&ID Drawing

Granted, the above information is a repeat of that shown on other project documents. While repeating the same information on several different documents is often discouraged, as we have mentioned before, since it is difficult to keep all the data up-to-date, the convenience and even the requirement of having device data on the loop diagram outweighs the down side of keeping the information current in multiple locations. One way to manage the data is to maintain it all in an electronic database, see Chapter 3 and Figure 3-3. The individual documents, including the loop diagram, can then draw the data from this master source.

Loop Diagrams during Design and Construction

Loop Diagrams provide a means for checking the completeness and accuracy of the measurement and controls (i.e., process control) design package. In other words, the process control design package is probably complete when all the information needed to complete the Loop Diagram is available. Loop Diagrams are useful during construction, and depending on the other documents prepared for the project, Loop Diagrams may be the primary construction document. Since a proper Loop Diagram is a complete representation of the interconnections of a process control loop, it makes make a great installation document. (Frequently, construction progress is measured by the number of "loops" completed.)

One of the last tasks of the control system group prior to startup of a new plant or process line is to check out the entire system to verify that the devices were installed correctly and ensure that the control system is ready for operation. These tasks may be accomplished with a loop-by-loop confirmation. These confirmations are called many things: loop checks, process system inspections, construction acceptance checks, or, in many homes, "Why daddy or mommy, didn't get home on time." All the necessary information to perform loop checks may be available on other design documents. However, it is much more logical, economical and convenient for the control system design group to use Loop Diagrams for this task.

The end of construction is usually defined in the construction contract. The term "mechanical completion" may be used. This is a very broad term that needs, but rarely receives, detailed definition. However, a common and perhaps simplistic definition might be: construction is complete, power is turned on, utilities are in operation, but no process fluids have been introduced.

Under a contract with these terms, loop checks might confirm that all the tag numbered devices are ranged and interconnected, completely and correctly.

With the advent of digital instrumentation, range checking becomes quite easy since many field process measurement devices, or instruments, are quite variable. Digital field instruments are calibrated at the factory for broad ranges; the device then needs to be set for your specific application. For example a pressure transmitter might have the capability of reading from 0 to 500 psig for the 4-20 mA signal. You may want this instead to read from 0 to 100 psig so you have to set it up for that digitally. The range setup and verification can be done quite simply in the field with a handheld communicator.

Loop checks also confirm the presence and pressure of the instrument air supply. Electrical power is verified to be at the correct voltage level, properly connected and identified with labeling as to the source circuit and with wire or cable markers in accordance with your standards. The owner's operating personnel might be involved in operating the utilities and overseeing the power system during loop checks. The presence of process fluids can be simulated, particularly with intelligent or smart field devices. A handheld communicator that connects to many intelligent or smart devices for setting up the instrument often also includes the ability to simulate signals through the installed instrument, which simplifies and streamlines the loop check task. Of course, signals can be simulated even when you don't use intelligent or smart field instruments by connecting into the circuit with a separate signal generator. There are many of these readily available for purchase.

Other contracts define the end of construction in different terms, and owner personnel might perform the loop checks as part of the pre-startup activity. In any case, for a startup to go smoothly a complete set of loop checks should be done and Loop Diagrams are the best document to use to facilitate that task.

Loop Diagrams are also of value to the Operations and Maintenance control system personnel. They are invaluable when troubleshooting a malfunctioning loop, during periodic device calibration verification and preventive maintenance, and after any modification of the measurement and controls system.

You might be surprised to find a range of opinion on the value of Loop Diagrams. An increasing number of firms have made the Loop Diagram and the P&ID the only record drawings for all control system devices—no longer keeping other control system documents in their records. At the other extreme (and amazingly, in the authors' experience), there are some design managers or even owners who believe the cost of developing Loop Diagrams is a needless waste of money. Other owners see the need for Loop Diagrams but believe owner personnel, perhaps the engineers and technicians who will operate and maintain the plant, can produce these more cost-effectively. Still others use outside help to develop the Loop Diagrams. There are firms that specialize in developing Loop Diagrams.

Some of the major control system suppliers might have the capability to develop Loop Diagrams in a manner that suits them. Some distributed control systems have configuration software that documents the software in a "loop-like" way.

Leaving the development of Loop Diagrams to a construction contractor is, in the authors' experience, not a good idea. The beauty of the Loop Diagram is that it presents all the information associated with the hardware. A construction contractor is likely to limit their responsibility to only a segment of the complete work, which is natural. They are not likely to address information that they did not provide, so your Loop Diagrams will lack information. In the authors' opinion, and based on painful experience, it is far preferable to have the designer or the owner prepare the Loop Diagrams and have the contractor revise them as needed during construction.

The use of a common project database rather than the use of other drawings as information sources can simplify the task of developing Loop Diagrams. There are project database management software products available commercially. Many companies have software developed in-house by dedicated project team members that will produce Loop Diagrams or listings of the same data presented on Loop Diagrams. Some software has the capability of developing and drafting Loop Diagrams directly.

Loop Diagrams – A Look Ahead

The newer digital automation technologies, such as fieldbus, will call for significant changes in some documents used to define instrumentation and control systems. An article in *Control*, titled *An Engineering and Construction Firm Tackles Foundation Fieldbus*, suggests one possible change is to replace traditional Loop Diagrams with fieldbus segment diagrams.

The article states: "A divide no longer exists between field instrumentation and the control system as it has with DCSs. The control system and the field instrumentation must be engineered together. We call the end result of this combined engineering effort the Plant Automation System (PAS). Devices are effectively wired in parallel on a H1 segment, and they share a single port on a process controller's fieldbus communication module. A shielded pair Foundation fieldbus H1 segment communicates at 31.25 Kbps and can address as many as 32 fieldbus devices...Loop diagrams are replaced by fieldbus segment diagrams."

Fieldbus segment diagrams show all of the devices in a single segment and the electrical connection details, including shielding and grounding for all of the devices in the segment. The segment diagrams will serve the same purpose for the PAS as the Loop Diagrams presently do for pneumatic, electronic, and shared display, shared control systems.

Summary

In this chapter we have discussed Loop Diagrams and have discussed their use and some of the pros and cons involved. After P&IDs they are perhaps the most important document for the technicians and engineers responsible for keeping a process in operation. Properly developed, they tell anyone troubleshooting a problem everything they need to know about the components that make up a specific system.

The authors are advocates of developing Loop Diagrams as part of the project deliverables. The development of Loop Diagrams by the system designer and the use of them by a construction project startup or commissioning checkout team ensures a last check of the accuracy and completeness of the design. Due to the useful detail, they also can improve the quality of the construction bids since there is little guess work left for the contractor in how the system will be installed and connected.

If the Loop Diagrams are produced by the control system design group in a timely manner, they are available for loop checks as part of mechanical completion (as defined in the contract) and for troubleshooting during plant startup. They do constitute a significant expense. Offsetting this cost are the increased accuracy of the work and the usefulness of Loop Diagrams during construction and operation, where they will more than pay for themselves.

Installation Details and Location Plans

Installation Details

Installation details present the requirements for proper installation of instruments. These details establish the owner's expectations regarding the mounting and mechanical connection of an instrument. Unlike the schematic representation of a Loop Diagram, an installation detail shows the actual physical requirements of an installation; it may detail the very nuts and bolts needed.

Installation details show where the connection will be made on the piping, tank, vessel, chest or duct, providing the relative position of the instrument to the process connection and to the floor. Installation details can also be used to show how the instruments are to be insulated, winterized, or otherwise isolated from high or low temperatures.

There is no ISA standard that defines requirements for these details, nor is there a reference to installation details in the *ISA Dictionary*. However, that does not mean there is a lack of information on installing instruments. Libraries of installation details are established and maintained by many plant owners, maintenance technicians, instrument manufacturers, engineering contractors, instrument installation contractors and perhaps even by you. Of course, the libraries all differ in detail and nomenclature. Furthermore, most of the owners of the details are convinced that their set will give the best results. However, even without a standard definition for these drawings, many follow general guidelines.

The drawings will be used in the field, so an 8 ½" by 11" drawing size is optimal. It is easy to carry and to reference during installation. Some details are designed in 11" x 17" format but are reduced to the smaller size for handling, so the designer should use text and symbol sizes with the understanding that the printed drawing will be reduced to half size for use.

Bills of Materials

For control system documentation, bills of materials or material lists will be found primarily on installation details. Bills of materials or material lists are tables with item numbers that correspond to callouts in the body of the drawing. The tables may also include quantity counts, and will certainly include a description of the items. The more information you provide on the installation detail, the more likely the final installation will meet the owner's expectations on completion of the project.

The manufacturer and a part number are often included for clarity. Even if the listing is understood to be "or equal," listing the manufacturer and the specific part number is one way to ensure that the features you want are understood. To be sure you have included sufficient information, consider that the information you provide should be sufficient to purchase the components.

Exact counts for instrument tubing lengths are not normally provided, since the required lengths will vary. In the quantity column some generic callout will be used when the quantity is not known, for instance, "A/R" may be used for "as required." However, some site standards may call for the installation details to list a conservative, but "standard", length for tubing to support the development of material take-offs. The total amount of material required is arrived at by taking-off or adding the quantities of components from other project documents. It is best if the quantities are clearly noted as estimates, subject to verification by the installer.

As stated before, the bills of materials (BOM) are probably integral to the installation detail drawing itself; alternatively, the BOM listing might be on a separate sheet or in the written specifications. This way the item number appears on a number of sheets, but the description only appears once. This approach was particularly common prior to computer-aided design and drafting (CADD; now often written as CAD). The material list was typed on a separate sheet and attached to the hand drawn installation detail.

An installation detail depicts the installation of a single instrument, unlike the Loop Drawing, which shows all the devices associated with the loop number. Separate installation details are commonly used to show how a device is mounted, how it is connected to the process and how the electrical, hydraulic or pneumatic connection is made. Your facility may use separate installation details for an I/P transducer and for a control valve, a detail for a pressure transmitter and a different detail for mounting stands. Therefore, there could easily be five or more installation details associated with one loop: one for a pressure transmitter connection to the process, one for an instrument stand, one for a separately mounted I/P, one for a control valve without an I/P, and a fifth for an I/P mounted on a control valve. Figure 8-2 shows a very common installation detail depicting installation requirements for a remote mounted differential pressure transmitter in gas service, drawing ID-101.

There should be some link or cross-reference in your documentation between a device tag number and the associated installation detail. Two common approaches are to call out the installation details on the Instrument Index; the other is to issue a unique installation detail for each device tag number or to list the tag number to which the detail applies on the drawing. In our example with the "remote mounted differential pressure transmitter in gas service" instrument, drawing ID-101, the first method would yield a reference to "ID-101" in the Instrument Index for every d/p (differential pressure) cell of that type.

The second method of using a unique drawing for each tag number would yield many individual copies of the same drawing, ID-101, with each copy differing only by the referenced instrument tag number. There are advantages to either approach. The former calls for fewer drawings and it makes drawing revisions a lot less frequent and simpler to issue. It does, however, assume that the concerned parties are comfortable with using the Instrument Index to direct work performed by the installation contractor. The latter method might be less prone to misunderstandings, but to apply it there must be a system, an adequate budget and the willingness to handle the increased number of drawings.

Sometimes site standards call for all installation details for a single tag-marked device be included in one drawing. This unwieldy requirement may lessen the standardization advantage to using installation details, since many more individual drawings will have to be prepared. Also, and possibly more importantly, it is difficult to fit all the mounting, installation, power and wiring connection information on a single sheet. One solution is to reference mounting and wiring details on the primary installation detail so only the primary detail needs to be issued or referenced to the device tag number. Nonetheless, complex or special instrument installations may require preparing a few dedicated and unique installation details, but this should not occur often.

Facilities with multiple processes may have different installation material requirements. In this situation a written instrument piping material specification can support the installation details rather than attempting to cover all the material combinations on the drawing. The different processes are then cross-referenced to the appropriate materials listed in the specification. This is similar to the system used for piping specifications. On less complex projects, the instrument piping specification may simplify the material callouts by standardizing on the "worst case" material specification, perhaps 316L low carbon stainless steel tubing and stainless steel fittings.

The use of a materials list allows the installer to make easy and accurate material take-offs, which are then used for material requisitions and purchase. The materials list also enables warehouse personnel to pull and bag the materials necessary for an instrument installation. The instrument installer can then pick up a bag containing all the materials necessary to install a particular device. And you thought one-stop shopping only applied to the supermarket!

Many plants and construction contractors have work rules assigning a responsible craft for specific work tasks, notably the mechanical installation—the pipe fitters, and the electrical installation—the electricians. To prevent conflicts, the process industry generally tends to avoid showing both electrical and mechanical installation details on the same drawing. Furthermore, having a third separate detail for installing the device allows assigning that work to either contractor, depending on local work agreements.

Types of Installation Details

There are three general types of installation details:

(Note: the Type 1, 2, and 3 classifications used below are merely for clarification. The type number is not used elsewhere.)

Type 1 – Inter-discipline Design (Design Coordination): Developed to transfer information between the design teams, notably between the control system designer and the piping designer. These details can be sketches or formal drawings. These design coordination documents show installation requirements for control systems, which are then incorporated into the piping design drawings. The design coordination documents are not useful to provide to the installation contractor, since by the time the installation contractor is working, the piping design is complete, the spool drawings have been issued, and the piping has been fabricated. Spool drawings are developed so that welded portions of a pipeline complete with fittings and connections can be prefabricated instead of being fabricated in place. Pipe spools save time and money. Once developed, this type of installation detail should become part of the company or plant design standard for use on all subsequent projects. Some typical items covered might include:

- Upstream and downstream straight run pipe length rules for flow (flow rate measurement) elements—X diameters upstream and Y diameters down, depending on the piping configuration.
- Magnetic flowmeter installed so that it is always flooded, and flow is either horizontal or vertically upward.
- Piping connections for thermowells; small pipe sizes will swage to 3" or 4", or make use of elbows and tees.
- "Root valve" (the first isolation valve at the process) requirements for pressure connections, perhaps a 6" long pipe nipple ending with a ½" diameter gate valve.
- Allowable location of taps in piping: top of pipe, side, bottom, anywhere on an arc from 10 degrees to 350 degrees, etc.
- Flange mounted level transmitter connection—size, flange rating, who furnishes isolation valve, etc.

Type 2 – Mechanical Installation: Shows how the instrument is to be mounted and connected to the process. These installation details are usually the responsibility of the control system design team, and they cover the installation from the root valve to the instrument itself. It is on these details that the tubing requirements are established—including the slope, tubing size, and relative position of the transmitter. These details also address the acceptable use of pneumatic components such as flexible hose for air and signal connections to a control valve.

Type 3 – Electrical Installation: Shows how the electrical connections are made. Either the electrical installation details are the responsibility of the electrical design team or the electrical design team is integrally involved in their development. The electrical design team ensures that the instrument installation is in concert with the balance of the electrical work, and also ensures that the installations meet local and national electrical codes. Information included in these details includes the use of flexible conduit, cable grips with open cable drops to instruments, handling pigtail wiring for instruments, junction box mounting, and, most importantly, conduit and wire tagging requirements.

Kept in Library

Installation details should be maintained in a library of standard details for your facility. The library will build over time as new devices are specified that call for new installation requirements.

Installation details are the responsibility of the control systems design team or maintenance group. In a perfect world, your facility should have installation details covering every type of instrument or control device in your facility. That being said, sometimes the manufacturer's installation detail will be used instead of a project-generated detail. However, a bit of prudence must be applied when using a manufacturer's detail by itself. The manufacturer's detail may not cover the complete installation or, more frequently, the manufacturer's detail will confuse an installer by providing generic information or information on options not used for that installation.

A complete set of installation details is generally developed as part of a project design by the control system and electrical designers. Of course, there are exceptions to this approach.

On occasion, an operating facility will subcontract the control system installation to an instrument installation contractor. The project team should ensure that the contractor's installation details will meet the site standards. This verification should, of course, take place prior to signing the installation contract, to prevent embarrassing and expensive change orders. Some instrument manufacturers also have the capability to undertake some or all of an installation contract. In both of these instances, developing or updating the site installation details might be included as part of the installation contract.

There are certainly more variations on installation details than have been covered here. The authors have seen increasing use of digital photos to present preferred installations, complete with overwritten notes and leader arrows to significant points. At least one oil industry plant does without any installation details for all in-house projects. They add some installation information to the Loop Diagram and use this for installation directions. Their installers have the skill necessary for a satisfactory installation.

Needed for Bidding

If the control system construction contract is issued for bid, installation details are an important bid item when soliciting construction quotes. Some projects have had to deal with contract change orders because, during the bid process, an owner provided out-of-date or non-representative installation details. The construction team later discovered that the actual installation requirements were much different from the basis for their quote.

Figure 8-1 is an example of a Type 1 installation detail developed to transfer information from one discipline to another during the project design phase. These details may have been developed as company or project standards, perhaps long before the present project was started.

Figure 8-1 shows thermowell installation requirements to provide accurate process fluid temperature readings. These installations provide a self-flushing thermowell installation to minimize the buildup of deposits on the thermowell protrusion into the process stream. This buildup would, of course, affect the accuracy and/or speed of response of the temperature reading.

Figure 8-2 is an example of a Type 2 installation detail. Type 2 details are usually defined, developed or selected from an existing library of standard details. They are the responsibility of the control system designers and are generated specifically for all devices in the project. They are issued to the installation contractor for use during construction.

Figure 8-3 is an example of a Type 3 installation detail. It shows the general-purpose conduit installation for an instrument or control component. In this instance, the detail was developed and is maintained by the electrical design group. Note the drawing number, ELEC-001.

Figure 8-1: Installation Detail, Type 1 – Thermowell

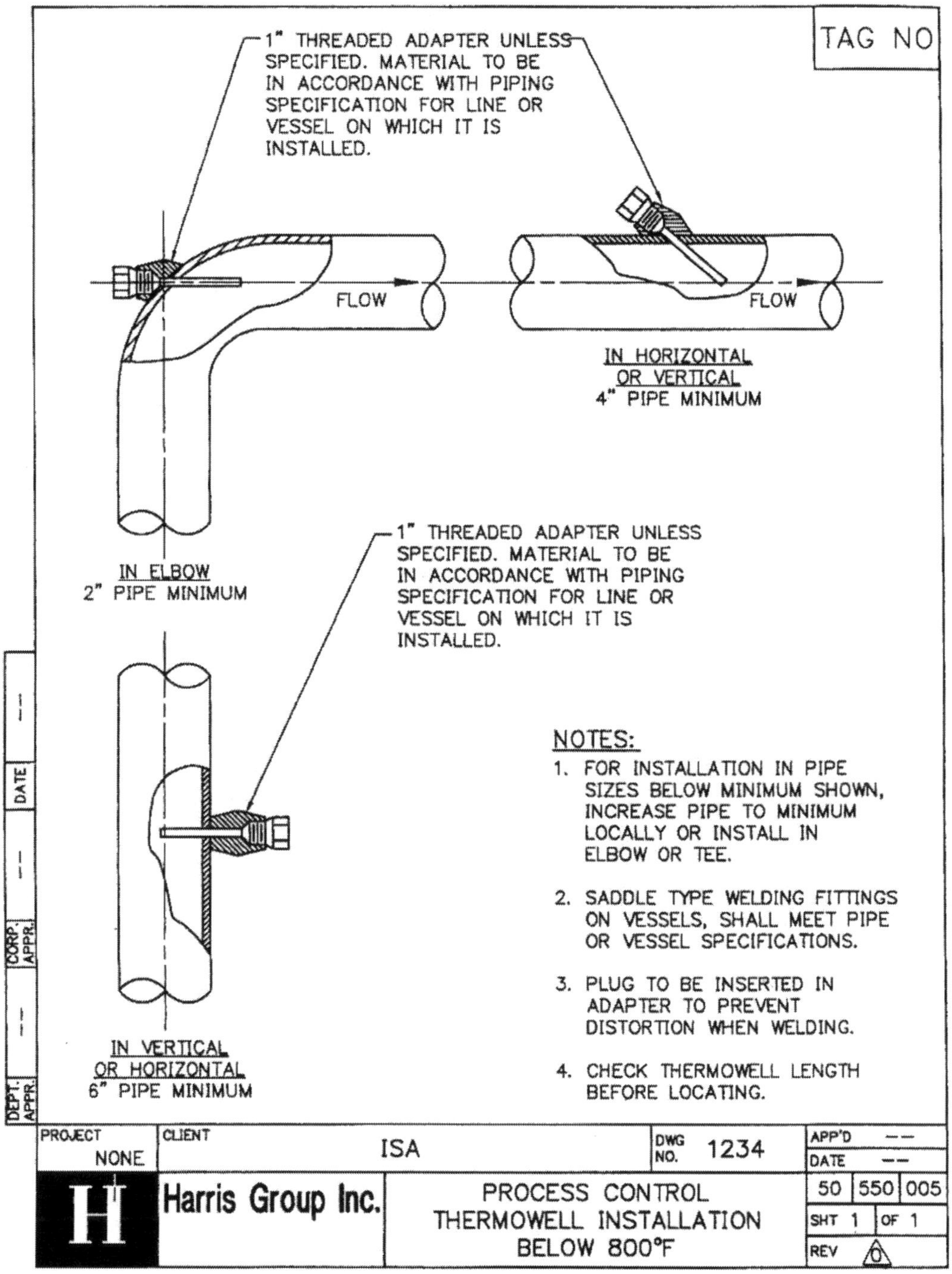

Location Plans

Location plans are orthographic drawings that show where control system components are to be installed. They can be important tools for training maintenance and operations staff; in the real world, when the control system flags a component as failed, it is not always easy to find it. The location plan

Figure 8-2: Installation Detail, Type 2 – Flow Transmitter

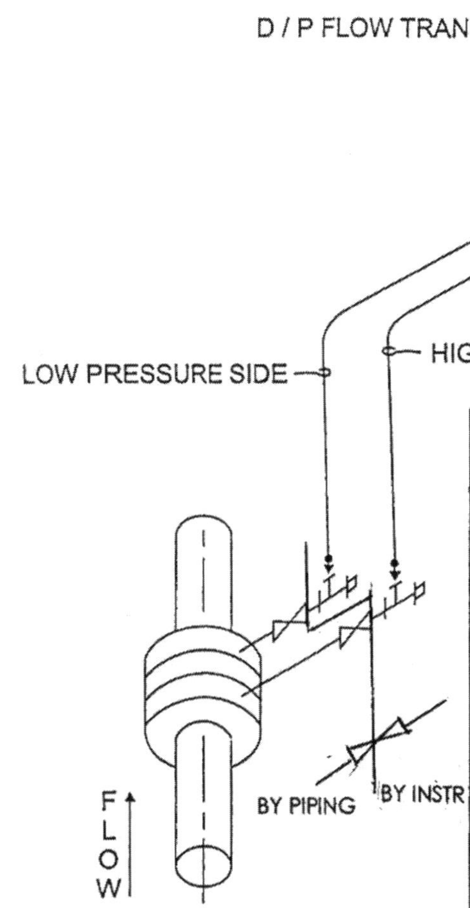

MATERIAL LIST		
QUANTITY	DESCRIPTION	MAIL
2	1/2 X 3" THRD. NIP	SHC 4C CS
2	1/2" THRD. TEE	↓
2	1/2" THRD. BAR STICK	↓
4	1/2 X 1/2" MALE CONNECTOR	316 SS
50 ft.	1/2" TUBING .027 WALL	316 SS
1	THREE VALVE MANIFOLD	316 SS

REV	DESCRIPTION	DATE	DR BY	APP BY
0	ISSUED FOR CONSTRUCTION	10/15/91	FAM	JAR

ISA COURSE FG15
D/P FLOW TRANSMITTER GAS SERVICE - REMOTE MOUNTED
INSTALLATION DETAIL
DRG ID - 101

can save a fair bit of time in searching. Location plan drawings are also useful when you are getting a bid from a contractor on the needed work, and they are almost invaluable during construction.

There is no ISA standard that details the requirements for a location plan. The *ISA Dictionary* contains no reference to them. The authors' definition, for the purpose of this book, is: "Location plans show the field control system components—the transmitters, control valves, transducers, local panels, junction boxes, termination points for field I/O, etc." A general rule is to show anything

Figure 8-3: Installation Detail, Type 3 – Conduit

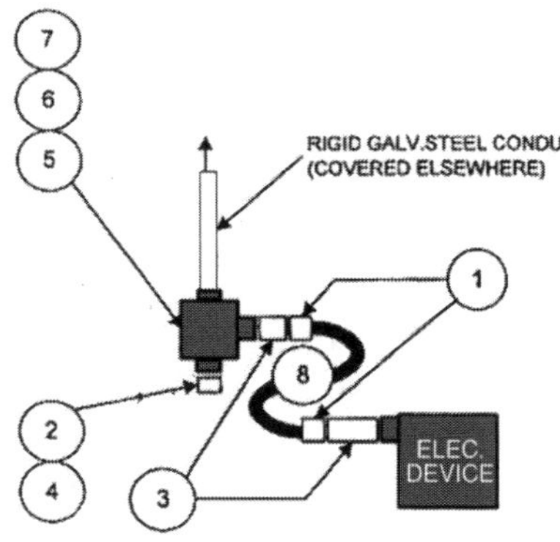

BILL OF MATERIAL				
ITEM	QTY	U/I	DESCRIPTION	PART NO.
1	2	EA	COUPLING,LIQ-TITE,FLEXIBLE	C-H LNM75
2	1	EA	'DRAIN,CONDUIT	C-H ECD15
3	2	EA	NIPPLE,CONDUIT,3/4" RIGID STL	FAB
4	1	EA	REDUCER,CONDUIT,3/4"X1/2"	C-H RE21
5	1	EA	COVER,CONDULET, FORM7	C-H 270
6	1	EA	GASKET,CONDULET,FORM7	C-H GASK572
7	1	EA	CONDULET, TEE,3/4",FORM7	C-H T27
8	2	FT	FLEX, LIQUIDTITE, 3/4"	VAR
9				
10				
11				
c				
13				
14				
15				
16				
17				
18				
19				
20				
21				

REVISION HISTORY			
REV	DATE	DESCRIPTION	INITIALS
A			

NOTES

TITLE	DETAIL NUMBER	REVISION
INSTRUMENT INSTALLATION DETAIL GEN'L PURPOSE FLEX IBLE CONDUIT TERMINATION DETAIL	ELEC-001	A

From Successful Instrumentation and Control Systems Design, by Michael D. Whitt. ISA 2004

the control system group "owns," meaning devices the group installs or interconnects with wiring or tubing.

In a location plan all field control system components are shown using simple notation, perhaps only the tag number and, if company or site standards call for it, the elevation of the device. The ubiquitous ISA instrument bubble is frequently used. The bubble is less suitable for densely packed drawings; it uses up a lot of space. When the location plans have to show many devices in small space, one effective solution is to use a simple text string for the tag. In this case, the control system device is located with a solid symbol like a small square. A leader line connects the solid symbol to the text string.

Location plans use, as their base, orthographic scale drawings of the facility. Normally, these scale drawing backgrounds are developed by other design groups, typically by the architectural, mechanical, or piping groups. This works well from a scheduling standpoint as the other groups will be, or should be, essentially finished with their design efforts by the time the control system designers start recording the locations of instruments. These locations are sometimes added to the electrical plans, and sometimes the control systems group prepares its own location plan. The quantity of instruments and the density of the electrical plans will dictate which system is used. There must be close coordination between the two groups to make efficient use of the electrical "backbone"—the cable tray and conduit runs. All parties should agree to the content and scope of the location plans before the design effort commences, since there are many possible variations.

Location plans work well using a fadeout piping drawing or, better yet, a fadeout equipment layout drawing. In a fadeout drawing, details that are not necessary for the location plan are shown using very light (faded) lines. This is quite easy to accomplish using CADD or, for the old timers, using a clever reproduction facility, printing screened drawings onto Mylar drawing stock. The information added for device locations is drawn in normal line weight so it will show clearly when contrasted with the fadeout background.

Figure 8-4: Location Plan, Approach A

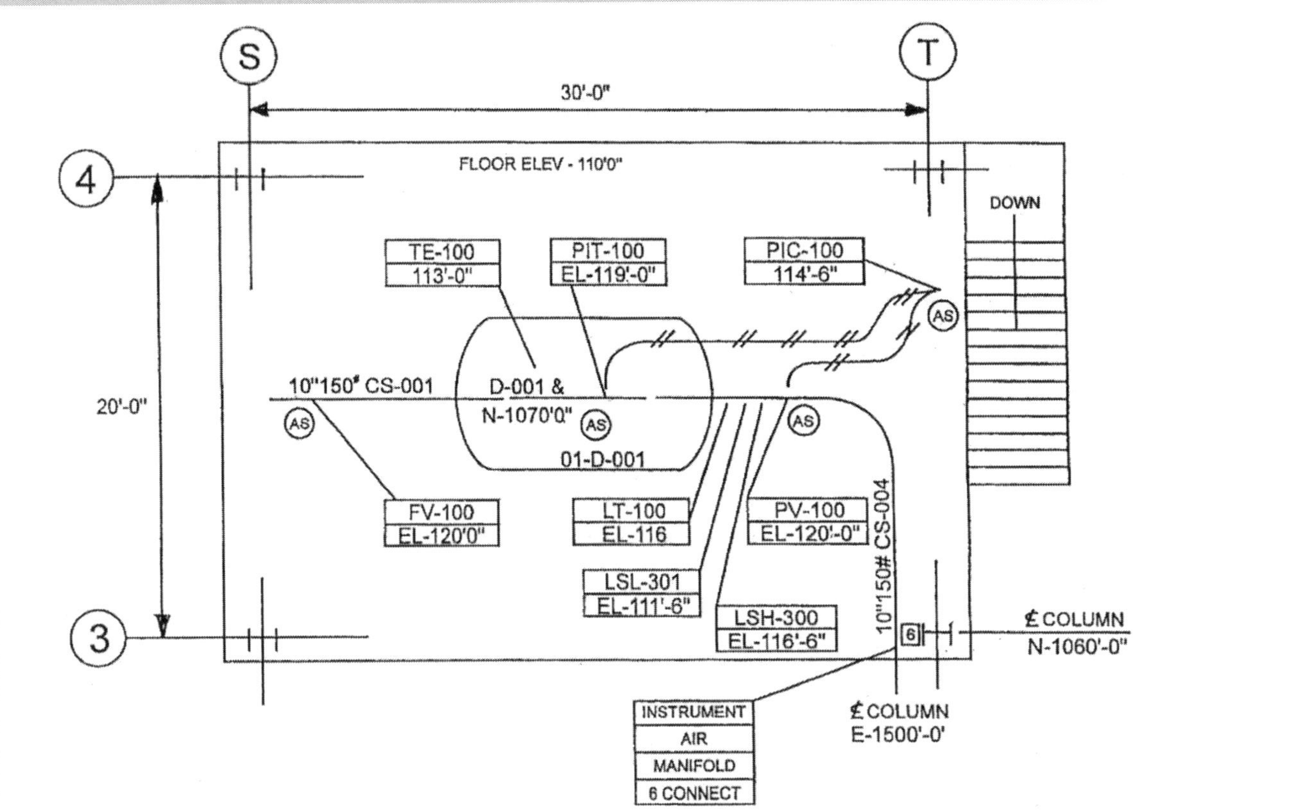

Figure 8-4: Location Plan, Approach A, is typical of one type of location plan. This approach locates devices orthographically and, sometimes, the elevations are indicated. This example uses some of the devices as shown on the example P&ID from Chapter 2, Figure 2-21. The location plan drawing shows the location of the instrument air manifold and defines the transition point between the instrument air distribution header and an instrument air branch header. The transition is important since at our example facility it will be the boundary between work by the piping group—the instrument air header, and the instrument mechanical contractor—the instrument air branch piping.

The plan also shows the devices connecting to an instrument air supply and shows in sketch format the necessary pneumatic transmission lines for loop P-100. The location plan uses the plant identification grid and rows and columns, including the dimension callouts. Therefore, grid locations for all the devices may be developed from this location plan. Location plans are useful when assigning individual instruments to marshalling or I/O panels; the designers can assign instruments to the nearest panel.

Grid References

Having accurate instrument locations can be very helpful to maintenance personnel, to the installation contractor and to the design staff. In fact, having locations shown in the Instrument List is a requirement at some facilities. This location information can be time consuming to acquire, so it can be expensive information. Making matters worse, for those who are working on a construction project, defining locations normally occurs relatively late in the project, when budgets are tight.

The question always arises: do you call out the nearest column intersection when defining a device location? Define *"nearest"*, because sometimes it isn't obvious which column intersection is closest. Using this method, the device could be in any direction relative to the intersection, but within half the distance to the next column. Alternatively, do you take the time to estimate the distance to the two nearest columns for a more exact map? In our opinion, an exact location in the list is rarely important. It is sufficient to simply identify the location within 20 feet or so; therefore, the column intersection is usually sufficient. There is little need to subdivide the column callouts. Location at column DD-21 is sufficient, rather than fine tuning the location to DD.2-21.4.

Next, you need to decide which intersection to use. We have had the most success with establishing early in the project that the "upper left" intersection, or more technically, using the northwest intersection as the callout. This way, the information is easy to figure out, easy to maintain and everyone knows how to use the information, since the "northwest" intersection rule is easy to remember.

The location plan for this project does not show the routing of the air supply tubing. The tubing for this project will be field-run and supported by control system construction personnel.

Many location plans do not show conduit runs. It is common for construction contractors to establish the conduit routing in the field so there is little value in adding it to the drawing. The level of detail for the electrical portion of the instrument location plan should mimic that provided on the electrical plan drawings. It may be useful to show major electrical raceways like multiple conduit runs and cable trays above a certain size, perhaps greater than 6". The space needed for the trays can then be coordinated with other disciplines. It is arguably better to have one person or team run all the raceway design, so it is often more efficient to include the control systems tray and main conduit runs in the electrical design package.

The example location plan, Figure 8-4, shows the location of TE-100, a thermocouple connected to a control board mounted indicator, TI-100. On this project, the location plan does not show the location for PG-1, since its location will be shown on piping drawings. No location is provided on the location plan for TG-1 since it will be shown on the vessel drawing of 01-D-001, the KO Drum. As you can see, the rules for locating devices must be worked out between the design disciplines and the owner before the design commences. Actually, they should be decided before the various design budgets are established.

Figure 8-5: Location Plan, Approach B

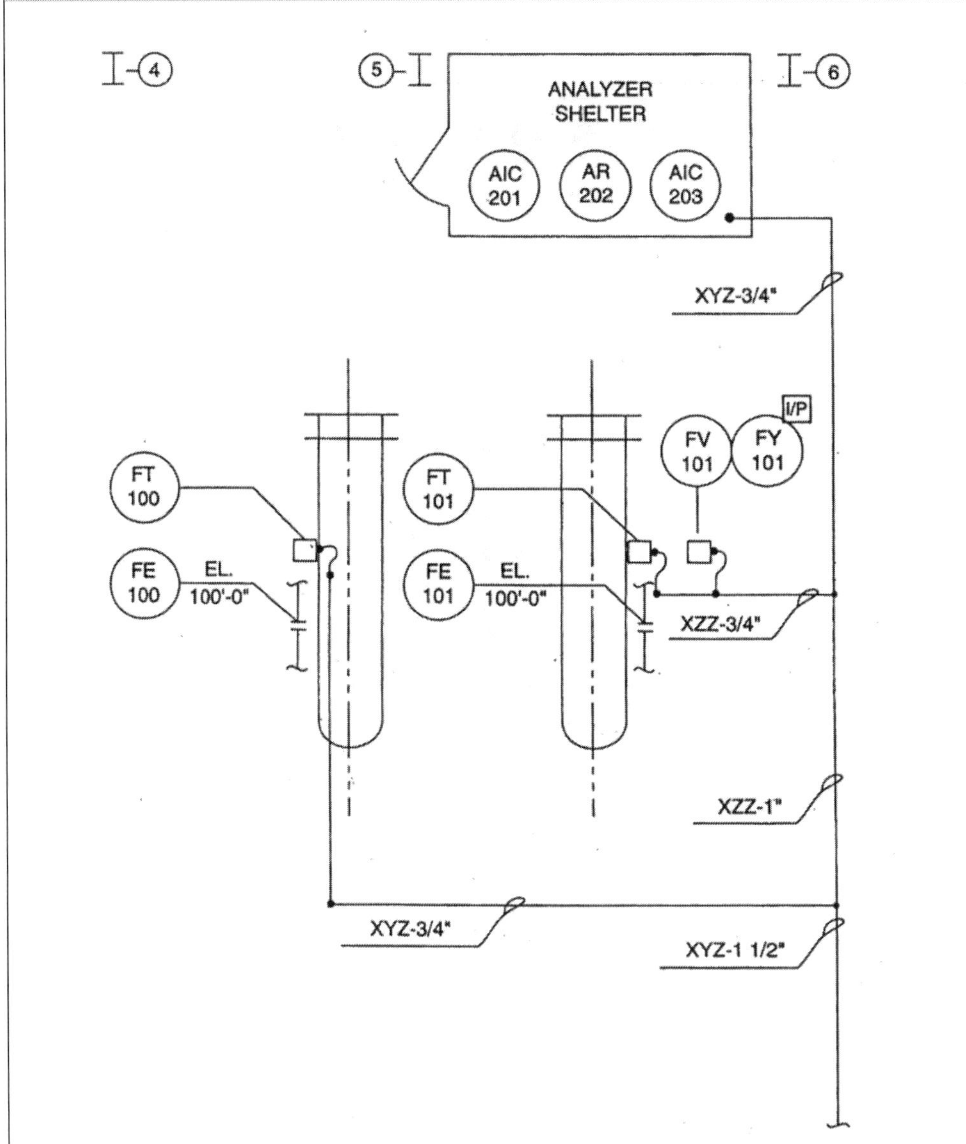

From Control System Documentation, by Raymond Mulley, ISA 1994

Figure 8-5, Location Plan Approach B, is drawn to a different set of project criteria. The P&ID on which it is based is not included in this book. This location plan shows the relevant column line, but not enough information is included to develop grid locations. The location plan uses standard bubbles or balloons and tag marks to identify the various devices. It shows the conduit size and layout for the electronic signal transmission system. Instrument and valve elevations are shown only when they differ from a project standard of 4' x 6" above the platform height. No information on instrument air supply or distribution is included.

Figure 8-6, Location Plan, Approach C, shows a location plan in development. All details are not included. This location plan differs substantially from Location Plans A and B. This location plan and the following description are taken from *Successful Instrumentation and Control Systems Design* by Michael D. Whitt, ISA, 2004.

Figure 8-6: Location Plan, Approach C

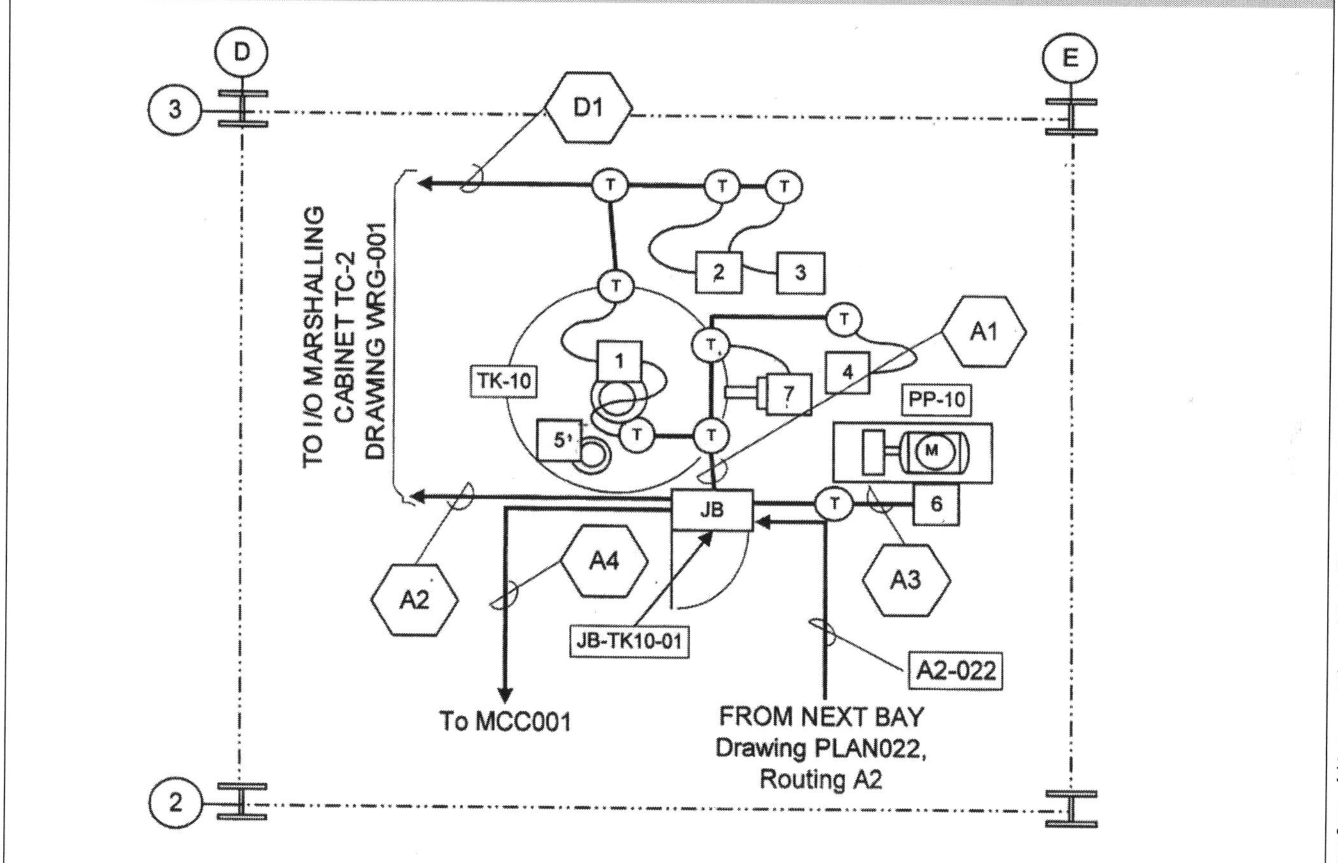

From Successful Instrumentation and Control Systems Design, by Michael D. Whitt. ISA 2004

"Instrument location plan drawings depict the location and identity of instrumentation and control equipment and provide conduit routing and cabling information. The location plan is typically spawned from an alignment equipment arrangement or rendered from a CADD three-dimensional piping arrangement. Once the background is completed showing the equipment arranged on the floor and enough steel detail to indicate location within the plant, the I&C designer layers on the instruments and shows conduit routings and contents. Junction boxes are depicted and cable routing schedules are produced. The graphics presentation can differ from one customer to the next depending on standards....

This drawing...shows our TK-10 product tank, a typical vessel with a level control system, fixed-speed discharge pump PP-10, and a recirculation system. The pump has a HOA switch and a contact closure to provide pump status to the computer. Instrument station boxes represent the instruments. The instrument station concept lets one box represent several instruments providing the instruments are physically grouped. In our simplified example one box represents one device. However, more than one signal may be generated by that device.

Conduit bodies are depicted with a circled letter. The letter indicates the general type of conduit fitting as denoting by the Crouse-Hinds nomenclature. The letter T indicates a tee-type fitting that could be the Crouse-Hinds T-style body or the explosion-proof GUAT-type fitting depending on the area classification....

Rigid conduit then carries the cabling to its destination. A component schedule describes the instruments represented by the station boxes and details the cabling that feeds them. Notice that items 4 and 5 each have two conduit connections, one for AC power and one for DC signal. Ultimately, the conduit size is determined and layered onto the drawing or listed in a cable schedule."

Detail Minimized

The basic location plan shows a minimum amount of detail. Tables are developed and added to the drawings as additional information is needed.

The control system devices are shown as numbered squares, 1 through 7. On a separate component schedule (not show), Item 1 is identified as a level transmitter, LT-TK10-10, and a limit switch, LSH-TK10-10.

Conduits are shown schematically and identified by lettered and numbered hexagons, D1 for DC circuits, and A1, A2, A3, and A4 for AC circuits. The cable origins and destinations are indicated.

A junction box is shown as a lettered (JB) rectangle with further identification shown with an adjacent rectangle (JB-TK10-01).

In a recent discussion, the authors asked a manager from a control system design firm how his company handles installation details and location plans.

He stated that they rarely, or never, provide installation details as part of their deliverables because their clients do not need or want them. His firm does produce location plans that differ from approach A, B, or C. Their location plans show the location of all tag-marked devices and their elevation on a background outline of major equipment with no interconnection information. Their clients use installation contractors or in-house installers who develop sufficient installation experience, coupled with information from other project documents, to install and connect the devices.

Location plans are not always included in typical design deliverables. If a detailed model is built or developed by graphical means for all or part of a plant, the location plan may be unnecessary. Other design documents and even the layout of the facility may preclude the need to define device locations and interconnection information. All this being said, it is normally very helpful to provide some form of an overall view of the device layout during construction. It assists field construction supervisors with setting crew size and work schedules, and confirms the availability of installation material.

Summary

The content of your drawing set must meet your facility's requirements. There is no universally correct answer to the use of installation details and location plans or their content, since each facility operates under diverse physical layout, operating, maintenance, design, construction and even regulatory conditions.

Some control system design packages include installation details and location plans that locate instruments and show instrument air details but do not show any conduit layouts. Some design packages include the location and elevation of pressure gauges and thermometers and others do not. Some design packages include a location plan that shows the locations and elevations of only the devices that require an electrical/electronic connection. Still other design packages include a location plan that shows the location and elevation of all tag-marked devices, but does not show any instrument air detail or conduit information.

At a recent technical conference the authors expressed their feeling that location plans were slowly falling out of favor because, as facilities age, the location plans become increasingly inaccurate or even get lost once construction has ended and plants go into operation. A number of operating personnel, on hearing this statement, disagreed. To them the location plan was one of the most valuable control system documents they had. Since the alternative would be to find a faulty field device by searching through many drawings, they valued the location plan as a time- and money-saving document.

CHAPTER NINE

Drawings, Title Blocks & Revisions

The medium upon which we draw has changed dramatically over the past 40 years. What were once ink lines on linen are now bytes on magnetic media. Drafting boards are now desktop computers. Despite these changes, some requirements remain—people looking at a drawing want to know at a glance what it shows, if it changed since the last time they looked at it, and who made the drawing different from what they expected. We also still want the physical copy of the drawing to fit nicely in a file drawer. These requirements are seemingly simple, but are often not met as different designers are involved with generating the drawings that define your facility.

To increase the likelihood that some of these wishes will be met, your company will probably have already developed standards for your facility as has been suggested this many times in this book. Included in those standards will be the most basic elements of a drawing: the border, the title block, the revision process and revision documentation, and even how to number and title the drawing itself, including the electronic file.

The layout of a drawing has not changed much over the years. Several features characterize a drawing template:

- Drawing Size
- Border
- Title Block
- Revision History
- References, Scale and Notes

A template is what the drawing looks like before someone adds content to it—for the old timers, it was all a new, blank drawing had on it when you pulled it out of the drawer in the drafting room, before anyone drew on it. Today, a computer-aided design (CAD) program should be set up to generate a template automatically.

Drawing Size

Drawing sizes in the United States are defined in ANSI (American National Standards Institute)/ASME (American Society of Mechanical Engineers) Y14.1 - 2005 *Decimal Inch Drawing Sheet Size and Format*. In it you will find the dimensions and features that comprise an effective drawing. The market also establishes sheet size; only certain sizes are typically used within an industry.

The standard ASME drawing sizes are identified with a single letter. A summary of the available sizes used in engineering is given in Figure 9-1.

Figure 9-1: ASME Document Sizes for Engineering Drawings

	ASME Letter	Size	Margin	Typical Uses
Notebook Size	A	8-1/2" x 11"	H-0.38" V-0.25"	Specification Forms Installation Details
	B	11" x 17"	H-0.38" V-0.62"	Loop Diagrams Installation Details
Desktop Size	C	17" x 22"	H-0.75" V-0.50"	Same as Size D and E, Logic Diagrams, Interconnection Diagrams
Full Size or "Bed sheet"	D	22" x 34"	H-0.50" V-1.00"	Process Flow Diagrams Piping and Instrumentation Drawings
	E	34" x 44"	H-1.00" V-0.50"	Orthographic or Scale Drawings Panel Layout and Wiring Same as Size C
Other sizes, less common, include F thru K and Roll Sizes (For Architectural Document sizes, see Appendix D.)				

From ANSI/ASME Y14.1

In some ways, the function of the drawing determines the size. Drawings attached to text specifications are normally A size so they can be easily bound to the parent document. Drawings used for troubleshooting and maintenance will be carried into the field, so it is useful when they easily fold to fit in a pocket. Thus, Loop Diagrams and Installation Details tend to be B or A size drawings. Drawings that are scaled or orthographic representations are normally D or E size since, at typical scales, the area available on the drawing is representative of a reasonable area of the facility.

The facility owner or the design project management team usually establishes the allowable drawing sizes. It may sound trivial, but having a mixture of C, D and E size drawings presents a problem for drawing storage files, drawing racks, and even for paper stock in the copiers.

The C and D size drawings have become somewhat more common with the transition of office work areas from drafting tables to standard desks. A 17" x 22" (C size) or 22" x 34" (D size) drawing fits nicely on a desktop, yet the available area on the drawing is large enough for practical use. Many non-scaled drawings like P&IDs and Logic Diagrams are often developed in the larger D and E size format, but they are plotted and distributed as C size, or even B size, for convenience—not to mention the inherent savings in reproduction costs and space.

Reproduction costs are based on the area of the copy. Smaller is better as long as the drawing is legible. Orthographic and other drawings produced to scale are normally filed and distributed as full size drawings to maintain their "to scale" nature. Drawings that carry a PE (Professional Engineer) stamp are required by some state laws to be stamped and filed as full size drawings. Today, if reduced sized copies of stamped drawings are needed, they may be made as reduced size photocopies of the wet-stamped original, a process that can be expensive and time consuming. Some specialized reprographics firms have the equipment to reduce an E size drawing down to C size.

There are some projects, and therefore some facilities, that will incorporate multiple copies of A or B size drawings onto a single E size sheet for issue and storage. In the authors' opinion, this should be strongly discouraged. It defeats the key advantages afforded by smaller drawings; you lose the portability benefit and the simplicity of locating a single drawing from a coded drawing number. Having one drawing containing four Loop Diagrams makes it hard to relate the loop numbers to the drawing number. Locating the "right" drawing from a list becomes more difficult.

Non-standard Drawing Sizes

From a practical standpoint, a photocopier will likely only handle two paper sizes, A and B, and a plotter will be limited to drawings no larger than D or sometimes E. Drawings have to be made to fit those sizes. You may have seen, first hand, difficulties arising from having to handle non-standard drawing sizes. Have you ever experienced a printer jam while printing documents that came from another country, or possibly a manual downloaded off the Internet? The printer needs instruction on how to handle the unexpected paper size request. European ISO metric paper sizes are different than those used in the United States; for instance ISO size A4 is narrower and taller than 8-1/2" x 11" paper. The printer either won't accept it or the end result will be reduced to fit into the smaller border.

Figure 9-2 lists the European and the comparable U.S. sizes. Your printing software might make the conversion automatically, but the layout of the document will look odd. For the unprepared, working on an international project with offices in Europe and the U.S., these differences in document sizes can make for some annoying disruptions while the conversion tools or procedures are developed.

Figure 9-2: European Document Sizes

	ISO Letter	SIZE Width x Length mm (inches)	Like ASME	ASME Size Width x Length Inches
Notebook Size	A4	210 x 297 (8.3" x 11.7")	A	8-1/2" x 11"
	A3	297 x 420 (11.7" x 16.5")	B	11" x 17"
Desktop Size	A2	420 x 594 (16.5" x 23.4")	C	17" x 22"
Full Size or "Bed sheet"	A1	594 x 841 (23.4" x 33.1")	D	22" x 34"
	A0	841 x 1189 (33.1" x 46.8")	E	34" x 44"

From ASME Y14.1

Equipment suppliers are not driven by the same drawing rules as the facility in which the equipment is installed. Consequently, despite a facility's best efforts to maintain site drawing size standards, many different drawing sizes can be found on the vendor drawing stick files. This presents a problem in handling and filing. Vendor drawings may range from A size installation details to a scale drawing of a paper machine or boiler that can be 15 feet long or more!

The increasing use of electronic drawings can make drawing storage a bit easier. Vendor drawings are sometimes issued to the purchaser as standard CAD files, assuming that the two companies have the same or compatible systems. More typically, drawings can be exchanged using some universally readable software like Adobe Acrobat™. These are easy to store and read on a desktop computer, but there may be some challenges when a printout is needed. Is the right paper available? Will the third-party drawing plot on your system? They don't normally, due to different system setups. None of these problems are insurmountable, but they can be annoying.

Border

The drawing border is, of course, the drawing limit. However, more useful features are included in the template. For starters, the ASME standard, not to mention years of experience, calls for a margin of unused space around the drawing. The margin provides a sacrificial zone where day-to-day damage to the drawing's edges can occur without losing data.

The ASME margin sizes (see Figure 9-1) seem to be handled as optional recommendations; many variations on margin size can be experienced. The drawing border itself is typically a continuous bold rectangle that establishes the fixed outline of the drawing. It is the frame of the picture. Border dimensions ("size") are listed in Figure 9-1.

When drawings are printed using a plotter loaded with varying paper roll stock sizes, your standard CAD template should include short light lines in each corner showing where the drawing sheet should be cut so the final product is the correct size. These cut lines are normally about ½" or 1" outside the bold line defining the drawing limit. The cut lines correspond to where the edge of the paper should be when you pull it out of a drawer. A plotter doesn't normally plot to the edge of the paper, so this edge definition makes for tidy drawings.

A grid system can be added to any drawing to provide a cross-reference, similar to that provided on road map. (If you think about it, at some point soon, hardly anyone will know what "road map" means—old-timers will say, "Maps are those paper sheets that you see in pirate movies. Back in the olden days, before GPS systems, maps of roads were found in every car too.") Anyway, a grid system is more useful on larger drawings (C size or above) with more surface area and therefore more need to guide the user to a smaller area of interest. The grid appears adjacent to the drawing border, and is marked with short hash marks at intervals, every four inches or so. Resulting sections are then identified with sequential numbers and letters on the horizontal and vertical edges. When referring to the location of, say, a control valve on a P&ID, you could use the grid to provide the general area to search for the device, for instance "Grid C-9." The valve can be quickly located in the vertical C area and the horizontal 9 area. In reality the valve would be much easier to identify by its function description in the loop name, but there will be times when the grid comes in handy.

Title Block

The title block is the most familiar feature of a drawing, almost always appearing in the lower right corner of a drawing. Figure 9-3 shows a typical Title Block arrangement.

Figure 9-3: Typical Title Block

<table>
<tr><td rowspan="2">Design By</td><td rowspan="2">Date</td><td>Owner's Logo</td><td colspan="4">Facility</td></tr>
<tr><td>Funding No.</td><td colspan="4">Area</td></tr>
<tr><td rowspan="2">Drawn By</td><td rowspan="2">Date</td><td>Designer's Logo</td><td colspan="4">Drawing Type</td></tr>
<tr><td>Contract No.</td><td colspan="4"></td></tr>
<tr><td>Check By</td><td>Date</td><td rowspan="2">PE Stamp</td><td colspan="4">Drawing Title</td></tr>
<tr><td>Appr. By</td><td>Date</td><td>Scale</td><td>Size Letter</td><td>Drawing Number</td><td>Rev.</td></tr>
</table>

Drawing title blocks usually consist of multiple lines—as many as five, or as few as one. When multiple lines are used, each line should present the same information for every drawing within the facility. A typical title block line convention is provided in Figure 9-3; your facility's drawing standard should include a similar convention that meets your needs:

- **Facility** – Optional information, possibly the name of the plant site.
- **Area or Division** – Commonly provided subset of the plant site or entity furnishing or using the drawings.
- **Drawing Type** – The function or purpose of the drawing. This information may be encoded in the drawing number, but it is helpful to tell those using the drawing its purpose, even when that is self-evident. Here callouts like "Process Flow Diagram," "Piping and Instrumentation Diagram," "Loop Diagram" or "Location Plan" will be found.
- **Drawing Title** – This is really the primary title field of the Title Block. All too often, though, some less-than-helpful names make it into this field. For a P&ID, for example, the equipment depicted should be listed, rather than the less-than-helpful "Process Area, Sheet 1 of 3," which often happens. The title should explain to those reading a list of drawings exactly what the drawing contains. Not adding descriptive information is simply lazy and lousy customer service. For a Loop Diagram, the "Service" callout that appears in the Instrument List and on the Specification Forms may be used.

As with other drawing features, the layout of the title block really ought to be consistent for every drawing used in a facility, except for the standard equipment drawings generated by suppliers. This is true regardless of which designer made the drawing or when it was generated. Facility staff might have a hard time convincing contract designers to use the facility's template, but with perseverance such issues can (and should) be resolved to everyone's satisfaction.

Drawing Number, Sheet Callout, and Scale

The drawing number appears in the bottom right corner of a drawing, immediately to the left of the drawing revision number. Drawing numbers can be encoded with letters and numerals that define the drawing's function, which is helpful when arranging and filing drawings. The drawing number may include letters or number combinations to identify facility or site, the function area of the facility or the process to which the drawing applies, the type of drawing or the discipline whose work is depicted. Even after the coding, there will also be a sequential number. For orthographic drawings, the drawing number may also be encoded with the floor, grid section and elevation. There are as many drawing numbering schemes as there are facilities that have drawings.

For control system drawings, it is useful to allow for multiple sheets to be identified under the same basic drawing number; the base drawing number defines the drawing type and purpose. Sheet numbers are added under the base number to increase the space available so that all the information can be shown. For instance, when the loop number is the drawing number, more than one sheet might be needed to show all the components that make up the loop.

For construction projects, when all the drawings from all the disciplines are packaged as a set, individual sheet callouts, in the form "sheet N of Y" can be used to simplify assembly of the drawings in the correct order. Once the project is complete, though, the sequential number may no longer be needed. Since the drawing resides in your file system as a unique document, it should stand alone, easily located by its drawing number and the title.

The sheet callout should logically appear near the drawing number, The form is well-established: "Sheet X of Y." Occasionally, you will see just a sheet callout with no "of" quantity, but that is rare. It is often important to know how many sheets are included in the set. The sheet number can be included in the title block line itself when multiple sheets occur in a drawing template that has no provision for sheet numbering. The authors are aware of an example of doing this. New equipment was added to a P&ID, but there was insufficient room to add it on the existing sheet. P&IDs are supposed to appear in process order, so it was decided to split the "full" P&ID into two sheets, adding the additional equipment into the new space. The title of the P&ID then just had "Sheet 1 of 2" added to the text. This was awkward, but it worked.

For orthographic or section drawings the primary scale of the drawing is defined in a dedicated box on the title block. There can be instances on one drawing when a specific area is drawn to a different scale, but that special area scale would be defined in the section or detail title. For P&IDs, Loop Diagrams and other non-scaled drawings, the scale callout should say "None," or some similar statement.

ANSI/ASME Y14.1 calls for identification of the drawing size by adding a block showing the ASME size letter designation. This practice, in the authors' experience, is rare, but when drawings are filed by size in different flat file drawers, knowing the drawing you are looking for is a B size sheet might be helpful. If your drawings are stored electronically, as many are now, knowing the size may be less important. Furthermore, if your drawing standards are in order and the drawing numbering scheme is effective, adding a separate callout for drawing size becomes superfluous. Most facilities try to standardize on three sheet sizes, A, notebook size for installation details, B for loop diagrams, vessel spec sheets and motor starter schematics, and D or E for everything else.

Revision Number

The revision number is normally provided in the lower right-hand corner of the drawing, although some rely on the revision table alone to carry this information. If the revision callout is provided in the lower right, there should be a table elsewhere on the drawing detailing the revision history. This revision table is addressed later in the chapter.

Owner's Information

Identifying the facility owner is probably called for by the site drawing standards; however, that may be done by the incorporation of the company logo. Although it's likely to be a company management decision, the size of the logo should balance corporate pride against the value of drawing space. Big logos leave less space for the drawing! Many facilities require the incorporation of funding information in the title block—that is, listing the account number under which the project will be designed, constructed and installed; this is helpful during design and construction, but it is probably useless for operations.

Designer Information

Designer information includes both required information and information the designer would like to add. The single most important "required" information is the small square designed to fit the Professional Engineer's Stamp. State laws differ, but they often call for application of a PE "wet" stamp—that is, the inked imprint of a Professional Engineer's stamp, signed and dated, on all construction documents, as defined by the state. Some states may not allow printing an electronic image of the PE stamp, and many do not allow an electronic signature imprint. The "original" drawing has to be physically stamped, signed and dated by the engineer. This may change in the future with the advent of electronic signatures, so the person or group responsible for such things should consult the appropriate state regulatory agency.

Shared discipline drawings like a P&ID may carry two or more PE stamps—some combination of the process designer, piping design engineer, and control systems engineer.

The designer's company or division has some pride of ownership in their work in producing the drawing, so they may wish to add their logo as well. Some facility owners set limits on such logos by establishing a fixed space within the title block for that purpose. It is helpful to the design firm to have a space to add their contract number; it makes for simpler tracking of design time and filing of the drawings.

Initial Issuing Data

Another common feature of the title block is to record the initial issuing data for the drawing—that is, to identify who designed, drew, checked, supervised, and approved the drawing. Dates for each of these steps are recorded in the issuing data area of the drawing as well. This information is separate from that contained in the revision block; this area of the title block is reserved for the initial construction release of the drawing. Drawn By (drafted), Designed By, Checked By and Approved By are almost always shown; others are added as called for by project procedures. Due to space limitations, initials are most commonly used. Initials are usually written and dated by hand for the initial drawing issue. For CAD drawings, subsequent issues normally indicate the initials and dates in text, without signatures.

Other Template Features

Drawings change over time. Information is added and deleted as the process evolves. The evolution of the changes has to be recorded clearly so those using the drawings can quickly see what has changed. A revision block is a table used to record the revision activity on a particular drawing. Figure 9-4 shows a typical revision block.

Figure 9-4: Typical Revision Block

2	REMOVED T-109	CAM	11/25/03
1	ADDED P-101	CAM	01/03/02
REV.	DESCRIPTION	BY	DATE

ANSI/ASME Y14.1 calls for revision blocks to be located in the upper right corner of the drawing. The authors are more familiar with revision blocks next to the title block at the bottom of the drawing, but as long as consistency is maintained, either location is fine.

In Figure 9-4, the revision block is located at the bottom of the drawing next to the title block—hence the column callouts. Rev., Description, By and Date are at the bottom of the table, and the revision history proceeds upwards from the bottom. Note that Revision 1 is the first entry. The initial issue data was recorded in the title block. (See the previous section on Designer Information.)

There are a number of ways to handle the revisions when you run out of space in the title block. Handling revisions is easier when you use a CAD system. The revision record can roll over; the oldest revision line is erased and the newest added in that space, or the revisions are moved down a line to make room for the new revision and the superfluous line is erased. Sometimes extra revision blocks are added to the drawing wherever a clear space was found. This seems to the authors to be unnecessary; we can't see the point of knowing that Revision 1 was made back in 1965, although each company will have its own rules.

The description of the revision should be clear and as complete as space allows. You can understand why it is less than helpful to those reading the drawing to see the words "General Revision" in the description—they then have to play detective to discover what changed. Complex revisions associated with some large projects may even cross-reference the description column to another, more detailed text-based revision history document.

Drawing Change Notices

Sometimes, revisions are issued in smaller, more easily handled pieces; one company known to the authors calls them "Drawing Change Notices" (DCN). These A size documents are easy to email and fax and they allow the rapid publication and distribution of drawing revisions without the complexity of issuing a full size drawing. A DCN shows a small area of the drawing with the change incorporated. Usually, text is attached describing the change and listing approval data for the change—that is, approval signatures and change orders. The revision block on the drawing then just lists the DCNs that have been incorporated into the main drawing.

The organization required to handle a complex system like this is not suitable for every project. However, for complex projects involving many different disciplines, this has proven to be an effective system.

This discussion of DCNs is useful, if only for you to understand that a drawing is a tool to be used to meet your needs, and that the features of that drawing have value and should be maintained.

While it may not seem to you that preparing, revising, issuing (sending to concerned parties), filing for record, and even mounting on a stick file, a bunch of drawings would be complex and expensive, it most certainly is. A stick file, by the way, is a clamp that holds drawings in a group, which in turn mounts on a rack for easy access.

Revision Identification

There is normally insufficient space in the revision block to adequately describe changes to a drawing in text. In addition, text is time consuming to generate and is frequently misunderstood, so some other method of locating and highlighting changes is needed. Occasionally, the grid marks on the drawing are used to define zones where changes have been made. However, this leaves the challenge of describing the changes in words, and, accordingly, dedicating space on the drawing for that text.

Another, essentially ubiquitous (universal) practice is to outline the changes on the drawing itself. Since most lines on a drawing are straight or angular, a distinctive circle of connected arcs is often used. This is called a "revision cloud." The line weight is usually heavier than the secondary lines on the drawing but lighter than the primary lines; a line width of 0.08" has been used on CAD drawings with good results. See Figure 9-5 for an example of a CAD revision cloud.

Figure 9-5: Notes and Revision Cloud

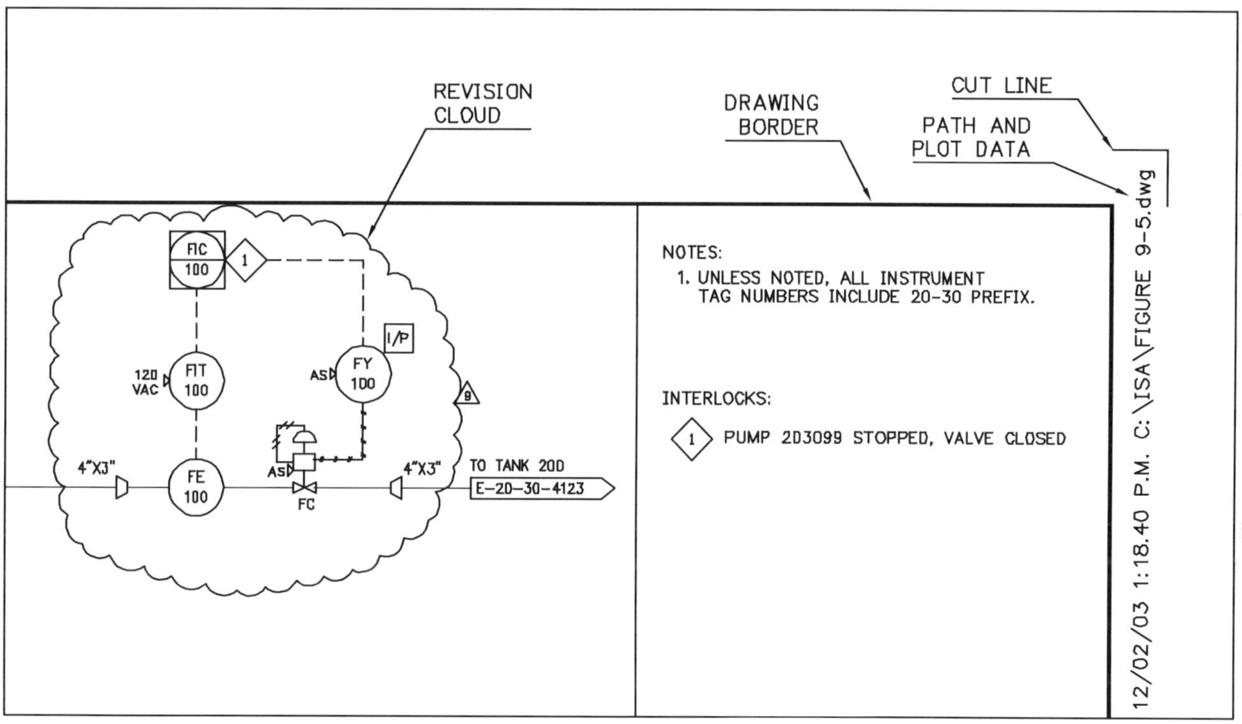

On manually created drawings, revision clouds were sometimes drawn with wax pencils on the reverse side of the drawing. This yielded a heavy line that was easily applied and removed. Wax pencils, or grease pencils as they are sometimes known, are available in colors so the clouds were easy to spot. CAD systems duplicate the procedure by placing the cloud on its own drawing "layer" so it is easy to locate and erase. The cloud is keyed to the revision block with the revision number circumscribed by a symbol, typically a triangle. A drawing may have many areas that changed, so there may be many independent revision clouds, each with a revision triangle showing the revision number.

When the drawing is issued again under a new revision, the old clouds are erased and new ones are installed, outlining the new changes. Sometimes local practice dictates leaving revision triangles behind as a memory jog that something in the area changed in the past. Others find the old triangles to be distracting clutter, and remove them.

Drawing Notes

Drawing notes, like other features, should be placed in the same location on all drawings for consistency. The notes are numbered sequentially and may be referenced in the body of the drawing. The notes section provides space for more lengthy descriptions that won't fit within the body of the drawing. Some companies have success with separate construction notes, that is, notes to the contractor that don't need to remain after installation is complete. These are removed before the drawings are issued for record.

Interlocks, if they are shown on the P&IDs, normally appear above or below the drawing notes, on the right hand side of the sheet. See Figure 9-5.

References Table

Documents referenced to build the drawing are listed in the references table. For example, Loop Diagrams may reference the P&ID containing the loop. Location plans are not normally referenced on a Loop Diagram, since that information is contained in the Instrument Index and the Index is in frequent use at the same time during a project that the location plan drawings are on everyone's desk. However, your procedures will vary to suit your company's work practices.

Logic Diagrams and possibly electrical elementary diagrams should reference the P&IDs that show the equipment for which the logic was written. It is sometimes helpful to list a vendor's equipment schematic if that is needed to understand that equipment's operation.

Plot Data

Some CAD drawings may include the very useful feature of automatically listing the file name and path to the drawing as well as the plot date in small text in the margins of the drawing. See Figure 9-5.

When you are under some project deadline and are madly printing out two hundred Loop Diagrams, it is sometimes helpful to know which copy of several is the latest one to print. Also, having the path identified may be useful if you have to find the drawing file later.

CHAPTER TEN

Role of Standards and Regulations

According to Webster's Dictionary, a standard is "something established by authority, custom, or general consent as a model or example." The control systems professional must be familiar with many standards. Some are mandatory, published in the form of regulations to be followed by law. Some standards are optional; after review by a responsible party, they may be found to be relevant to the activity being performed and therefore followed. If the responsible party should find them irrelevant, they would, of course, be ignored.

Mandatory Standards

In the United States, federal, state and local laws establish mandatory standards with the goal to keep everyone safe. These requirements have various names: codes, laws, regulations, requirements, etc. Examples include Food and Drug Administration (FDA) regulations, National Fire Protection Association (NFPA) Standard 70 (the National Electric Code—NEC), and Occupational Safety and Health Administration (OSHA) regulations. There are many others. The U.S. government manages about 50,000 mandatory standards for items as diverse as automobile air bags, children's aspirin and missile components.

Local law sometimes requires adherence to superseded revisions of national standards. For instance, your local town, city, or county building authority may stipulate the application of the National Electric Code (NEC) one or even more editions in the past. NEC editions are issued every four years. There is a cost to government agencies, plant owners, designers and contractors to use a new version of the NEC, a cost that local regulators may prefer to defer. Local laws may reference specific past revisions while lawmakers study the ramifications of the latest revision, and sometimes the implementation is not made because the latest edition does not change anything with particular local impact.

Adopting a new edition of the NEC into local codes takes time, so those who do project design work should check with the local building authority to determine the NEC implementation schedule for the project location.

Food and Drug Administration

The Code of Federal Regulations (CFR) lists the general and permanent rules of the Executive Departments and Agencies of the Federal Government. Title 21 of the CFR is reserved for the rules of the Food and Drug Administration (FDA). It is revised yearly. It can be searched directly at: http://www.gpoaccess.gov/cfr/index.html.

National Fire Protection Association (NFPA) Standard 70

The NFPA is an international non-profit organization with the mission to reduce the worldwide burden of fire and other hazards on the quality of life by providing and advocating consensus codes and standards, research, training, and education. Among the NFPA standards is Standard 70, the National Electric Code. The NEC, which carries the force of law because it is referenced in OSHA regulations (29 CFR 1910 Subpart S and 29 CFR 1926 Subpart K), defines the way we install safe electrical systems in the United States of America. The standard's scope is huge, covering all electrical installations including house wiring, petroleum refining plants and even explosives manufacturing. As described in detail in Chapter 4, the NEC contains a classification system for electrical construction in areas containing flammable materials. The classification system consists of three parts which, when taken together, define the hazards resulting from flammable materials. See Figure 10-1: Hazardous Area Classifications.

Figure 10-1: Hazardous Area Classification

The National Fire Protection Association (NFPA) issues NFPA 70, the National Electrical Code for areas where flammable materials are present. The National Electrical code defines an area classification system consisting of three parts:

Class

Group

Division

For Example: Class I Group D, Division 1

The first part, Class, identified with Roman numerals I, II, and III, denotes the nature of the hazardous material present: flammable gases or vapors, combustible dusts, or ignitable flyings, or fibers. The next part, Group, further specifies the hazardous material groupings using letters A through G; for example, A for acetylene, B for hydrogen, C for carbon monoxide, etc. There are, of course, many other hazardous materials included under each letter.

The third part, Division, defines the risk of the material being present. Division 1 is used where the hazard normally exists. Division 2 is used where the hazard exists in abnormal situations only. The resultant three-part area classification dictates the type of electrical installation necessary.

Occupational Safety and Health Administration

According to the *ISA Dictionary*, "In 1970 the U.S. Congress passed the Occupational Safety and Health Act which specified the requirements that employers must follow to guard against employees' illness and injury. The Occupational Safety and Health Administration (OSHA) enforces these requirements."

After the 1984 Bhopal chemical plant disaster in India, the U.S. Congress ordered OSHA to develop procedures to prevent similar accidents in the United States. OSHA issued its directives in 1992 as government document 29 CFR 1910.119, *Process Safety Management of Highly Hazardous Chemicals*. The directive requires that all installations handling certain hazardous materials develop a set of documents:

> "OSHA requires that employers develop and maintain information about process chemistry, technology, equipment and operating procedures and that this information be communicated to involved employees so that they understand the hazards of the process."

There are several paragraphs in the 29 CFR 1910.119 Process Safety Management directive listing the documents required. See Figure 10-2: 29 CFR 1910.119(d) Process Safety Information. Some of these documents require input from the plant control system group.

Figure 10-2: 29 CFR 1910.119 (d) Process Safety Information
Typical documents required

- Process flow diagrams
- Process chemical description
- Inventory amounts
- Material safety data sheets (define hazards)
- Safe operating limits
- Consequences of deviation
- Materials of construction
- Electrical classification drawings
- P&IDs
- Relief system design
- Ventilation system design
- Design codes employed
- Material and energy balances for processes built after 05-26-92

Paragraph 119 (d), Process Safety Information, lists the documents required for process safety. P&IDs (Piping and Instrument Diagrams) and process flow diagrams (PFDs) are among the listed documents. Of course, the P&IDs should reflect the current state of the plant and be easily understood by all operating and maintenance personnel. Predictably, this means to the authors that the P&ID symbols and identification methods should be developed and maintained with adherence to ISA-5.1.

Figure 10-3: 29 CFR 1910.119 (f) Operating Procedures Documents Required

- Procedures for operation, startup, and shutdown
- Operating limits
- Consequences of deviation
- Procedures for correcting deviation
- Safety and health considerations
- Properties and hazards of chemicals used
- Precautions to prevent exposure
- What to do if exposed
- Safe maintenance procedures
- Quality controls
- Control of inventory

Paragraph 119 (f), Operating Procedures, lists the documents required for safe operation. See Figure 10-3: 29 CFR 1910.119(f) Operating Procedures. Questions such as the following must be answered:

- How can the plant be operated safely?
- What are the operating limits?
- What happens if the limits are exceeded?
- How can process excursions be resolved?

Figure 10-4: 29 CFR 1910.119 (l) Management of Change Typical Documents Required

- Required for management of change to documentation, process chemicals, technology, equipment and facilities
- Typical documents required
  - Description of change
  - Temporary or permanent
  - Technical basis for change
  - Process hazards
  - Analysis of change
  - Resolutions of recommendations
  - Authorization requirements

Paragraph 119 (l), Management of Change requires companies to establish and implement written procedures to effectively and properly manage changes to documentation, process chemicals, technology, equipment and facilities. See Figure 10-4: 29 CFR1910.119(l) Management of Change. Change is defined as anything except the replacement in kind (exact replacement) of process devices and equipment. Any proposed change must be defined, analyzed, and explained to those concerned, and all questions must be answered. In other words, the change must be understood. Management must then approve proposed changes before they may be made.

> **OSHA Compliance Challenging**
>
> The OSHA regulations sound complete and complex, and some plants achieve compliance only reluctantly. An older plant the authors once visited included two areas that handled hazardous materials and several areas that did not. Some of the original information and documentation was no longer available or useful.
>
> For example, the plant P&IDs had been developed over a number of years, as process areas were added to the plant. Different engineering contractors designed the areas, each using their own standards. Different specialists had revised P&IDs over the years. There was no firm plant standard, so revisions did not always agree. Therefore, developing the whole documentation package was not easy.
>
> However, after successfully developing the OSHA documentation package for the required plant areas, plant management decided it was a valuable exercise. They proceeded to develop the whole documentation package for the rest of the plant, and determined it was very worthwhile.

Industry, Company, Government Entity Standards

Many, if not most, large firms have developed a full bookshelf of practices, specifications and standards that apply to their work. Some government agencies have done the same. Many of these large collections incorporate practices, specifications and standards developed by other organizations.

Consensus Standards

Consensus standards include recommended practices, standards and other documents developed by professional societies and industry organizations. The standards developed by ISA are the ones most often used by control system practitioners.

ISA – The International Society of Automation

The Instrument Society of America (ISA) was founded in 1945 as a nonprofit educational organization by a small group of dedicated professionals interested in the rapidly emerging field of instrumentation. The society's name has changed, in several steps, to today's "ISA – The International Society of Automation." The abbreviation ISA has remained constant. The Society's intent—to serve as an authoritative technical resource for its members worldwide—has also remained constant.

ISA published its first standard in 1948. Since then ISA has developed more than 150 automation, instrumentation, and control system Standards, Recommended Practices and Technical Reports. The differences in these documents are defined by ISA as follows:

"***Standard:*** A document that embodies requirements (normative material) that if not followed, could directly affect safety, interchangeability, performance, or test results…."

"***Recommended Practice:*** A document that embodies recommendations (informative material) that are likely to change because of technological progress or user experience, or which must often be modified in use to accommodate specific needs or problems of the user of the document."

"***Technical Report:*** A document that embodies informative material."

To clarify further . . .

Perhaps a bit of clarification of these definitions is in order. "Normative" in this context means normal. For the word "normative," substitute "normal practice." For the word "informative," substitute "giving information." Then the three types of documents might be defined as follows:

Standard: Guidelines which follow normal practice on a specific subject. The guidelines are unlikely to change. A large number of unbiased experts have reviewed the document and agree that the stated guidelines are very important to control system personnel.

Recommended Practice: Information guidelines on a specific subject based on the best information available today. These guidelines are subject to change. A large number of unbiased experts have reviewed the document and have agreed that the guidelines are very important to those developing and using process control technology.

Technical Report: Information on a specific subject. A number of unbiased experts have reviewed the document and have agreed that the information is of interest.

ISA follows formal rigorous procedures to develop Standards, Recommended Practices and Technical Reports. These procedures ensure that the documents produced are accurate, balanced, and fair.

Many ISA Standards, Recommended Practices, and Technical Reports have also received American National Standards Institute (ANSI) approval. ANSI was established in 1918 as the American Engineering Standards Committee. Its purpose is to provide standards to improve the quality and methods of American industry. ANSI approval procedures are different, but are equally or more rigorous than those of the ISA. ANSI has accredited ISA to develop ANSI/ISA documents.

Figure 10-5: ISA-5 Documentation Series

- ANSI/ISA-5.1-2009 Instrumentation Symbols and Identification
- ISA-5.2-1976 (R1992) Binary Logic Diagrams for Process Operations
- ISA-5.3-1983 Graphic Symbols for Distributed Control/Shared Display Instrumentation, Logic and Computer Systems
- ISA-5.4-1991 Instrument Loop Diagrams
- ISA-5.5-1985 Graphic Symbols for Process Display
- ANSI/ISA-5.06.01-2007 Functional Requirements Documentation for Control Software Applications
- ISA-d5.07.01 Piping and Instrumentation Diagram Documentation Criteria (in development 2011)

Figure 10-5: ISA-5 Documentation Series lists the titles and numbers of the standards in the ISA5 series. Quotes from the ISA-5 documentation series are included in the previous chapters. The series includes the following:

- ANSI/ISA-5.1-2009 *Instrumentation Symbols and Identification* presents a comprehensive set of symbols and the identification methods used to designate measurement and control devices. The standard is suitable for use whenever reference to measurement and control instrumentation, control devices and functions, and software applications and functions is required for identification and symbolization. ISA-5.1 is the standard most often used worldwide for identifying measurement and control devices and systems on P&IDs.

- ISA-5.2-1976-(R1992) *Binary Logic Diagrams for Process Operations*. The purpose of this standard is to provide a method of logic diagramming of binary interlock (binary logic) and sequencing systems for the operation of processes in many industries.

- ISA-5.3-1983 *Graphic Symbols for Distributed Control/Shared Display Instrumentation, Logic and Computer Systems*, was developed to supplement ISA-5.1 for the identification of instruments and control devices when shared control-shared display systems were used. However, the key elements of ISA-5.3 are now included in ISA-5.1, and ISA-5.3 will be withdrawn.

- ISA-5.4-1991 *Instrument Loop Diagrams* supplements the material in ISA-5.1 relating to Loop Diagrams, including some additional symbols and six typical instrument Loop Diagrams—two each for pneumatic, electronic, and shared display control loops. One of each type includes the minimum requirements and the second includes the minimum plus optional items.

- ISA-5.5-1985 *Graphic Symbols for Process Displays* establishes a set of symbols used in process displays to depict processes and process equipment.
- ANSI/ISA-5.06.01-2007 *Functional Requirements Documentation for Control Software Applications*. The standard establishes functional requirement specifications for control software documentation. It provides techniques for documenting control system software and provides a basis for validation of run-time application software; the standard is especially helpful in bulk chemical, fermentation, biotechnology, specialty, and fine chemicals industries.
- ISA-d5.07.01, *Piping and Instrumentation Diagram Documentation Criteria* this standard when completed, is intended to provide recommendations and guidance for designing piping and instrumentation diagrams. The aim is to be of value to many different industries. The group developing the document has been organized and working for over a year at the time of publication of this book. Visit **www.isa.org/standards**.

Other ISA Standards that are useful in developing a design package include:

- ISA-20-1981 *Specification Forms for Process Measurement and Control Instruments, Primary Elements and Control Valves* includes Specification Forms and instructions for their use in defining many measurement and control instruments, primary elements and control valves.
- ISA-TR20.00.01-2006 *Specification Forms for Process Measurement and Control Instruments, Part 1: General Considerations* is an updated version of ISA-20-1981 and includes 51 totally new or technically updated Specification Forms. The TR means Technical Report, a designation that allows frequent revisions and updates without the protracted approval process associated with standards. At the time of this writing, this TR is a work in progress. For the latest information on Specification Forms see the ISA website, **http://www.isa.org**.
- ANSI/ISA-84.00.01-2004 Parts 1, 2, and 3 *Functional Safety: Safety Instrumented Systems for the Process Industry Sector.* The title of these standards explain their contents. Their intent is to define requirements for designing, manufacturing, installing, commissioning, testing, maintaining, and operating safety instrumented systems (SIS).
- ANSI/ISA-88.00.01-2010 *Batch Control Part 1: Models and Terminology*, shows relationships between batch control-related models and terminology.

ISA Standards, Recommended Practices, and Technical Reports cover more than 40 different subjects.

Other Consensus Standards

In addition to ISA, other organizations also develop consensus standards to guide the control system professional. These organizations include the American Petroleum Institute (API), the American Society of Mechanical Engineers (ASME), the National Electrical Manufacturers Association (NEMA), Process Industry Practices (PIP), and the Technical Association of the Pulp and Paper Industry (TAPPI), among others.

American Petroleum Institute (API)

The API is the primary oil and gas industry trade association in the United States, with a membership of about 400 organizations from all parts of the industry. API aims to provide information and aid to the oil and gas industry to improve efficiency and cost effectiveness, comply with governmental rules and regulations, safeguard health, improve safety, and protect the environment. API develops and distributes technical standards and other publications on oil and gas industry subjects such as:

Exploration and Production	Safety and Fire Protection
Petroleum Measurement	Storage Tanks
Marine Transportation	Valves
Marketing	Industry Training
Pipeline Transportation	Health, Environment & Safety
Refinery	Instrument Installation

ASME

Founded in 1880 as the American Society of Mechanical Engineers, today's ASME promotes the art, science and practice of mechanical and multidisciplinary engineering and allied sciences around the globe.

There are at least two ASME Standards that are useful when developing control system documents.

ASME Y14.1 *Decimal Inch Drawing Sheet Size and Format*—see Chapter 9.

ASME Y32.11 *Graphical Symbols for Process Flow Diagrams* includes a series of symbols that depict piping and equipment.

National Electrical Manufacturers Association (NEMA)

NEMA Standards Publications No. 250-2008 *Enclosures for Electrical Equipment (1000 Volts Maximum)* covers the classification and description of enclosures for electrical equipment and the environmental conditions each one can withstand. This includes information on NEMA enclosures types 1 through 13. Measurement and control devices and control panels are built to NEMA standards which define the environment in which they can be reliably and safely installed.

Process Industry Practices (PIP)

PIP is a coalition of process industry owners and engineering/construction contractors that has developed (and is continuing to develop) recommended practices for all areas of the engineering, procurement, and construction process.

Scientific Apparatus Makers Association, (SAMA)

SAMA no longer exists as an organization. Some years ago, SAMA developed PMC 22.1 *Functional Diagramming of Instrument and Control Systems*. PMC 22.1 is still in use in some industries, especially to document boiler control systems and functions. It details the function of the control system by a series of symbols, showing, for example, the modes of control in a controller, the proportional (gain), integral (reset), and derivative (rate). Functional Diagrams supplement, but do not replace, P&IDs.

At the time of their withdrawal, SAMA gave ISA permission to use the SAMA work. Portions of PMC 22.1 have been directly incorporated into the current edition of ISA-5.1.

Technical Association of the Pulp and Paper Industry (TAPPI)

TAPPI is a technical association founded in 1915. Over 20,000 individual members are involved in the pulp, paper, and converting industry. TAPPI has developed and revised Test Methods and Standard Practices for use in the industry.

Location-Specific Documents

These are the standards and other documents developed for a specific location or a specific project. They may be a simple document or they may be a very specific and detailed set of books. They may be as basic as the legend sheet that should (but is not always) included with every issue of a set of P&IDs. Some projects add additional information, including typical P&IDs and other sample documents.

International Standards Organizations

There are two international standards organizations that develop standards for worldwide use—the International Electrotechnical Commission (IEC) and the International Organization for Standardization (ISO).

International Electrotechnical Commission

IEC (http://www.iec.ch), founded in 1906, is a worldwide organization for standardization that is comprised of national electrotechnical committees. The object of the IEC is to promote international cooperation on all questions concerning standardization in the electrical and electronic fields. The IEC also publishes International Standards. IEC issued 441 publications in 2009, including 366 International Standards.

There are 59 national electrotechnical committees with full IEC membership and 21 national committees with associate membership. IEC headquarters are in Geneva, Switzerland. The United States member of IEC is ANSI (American National Standards Institute).

International Organization for Standardization (ISO)

ISO is one of the world's largest developers and publishers of international standards. ISO is a network of the national standards institutes of 163 countries, one member per country, with a Central Secretariat in Geneva, Switzerland that coordinates the system,

ISO develops voluntary standards that are very useful in facilitating international trade. For example, the ISO 9000 series are generic quality management standards, which confirm that a location has a system in place to maintain quality by managing its processes or activities. The procedures to confirm conformity to the standard have been developed by ISO together with the International Electrotechnical Commission (IEC). The conformity assessment, or outside audit, is carried out by other organizations.

ISO is a nongovernmental organization that forms a bridge between the public and private sectors. On the one hand, many of its member institutes are part of the governmental structure of their countries or are mandated by their government. On the other hand, other member institutes have their roots uniquely in the private sector, having been set up by national partnerships of industry associations.

Appendix A
Answers to Chapter 2 Exercise

Figure 2-19, Descriptions

Figure 2-20, Symbols

1. T
2. A
3. C
4. P
5. S
6. D
7. J
8. H
9. R
10. F
11. B
12. L
13. E
14. G
15. O
16. N
17. Q
18. I
19. K
20. M

Appendix B
Abbreviations

Abbreviation	Explanation
AC	Alternating Current
AI	Analog Input
ANSI	American National Standards Institute
AO	Analog Output
API	American Petroleum Institute
A/R	As Required
ARO	After Receipt of Order
ASA	American Standards Association (NOTE: Superceded by ANSI)
ASME	American Society of Mechanical Engineers
BPCS	Basic Process Control System
CAD	Computer-Aided Design or Drafting
CADD	Computer-Aided Design and Drafting
CE	European Community Mark, indicating compliance with European requirements
CS	Carbon Steel
CSA	Canadian Standards Association
DC	Direct Current
DCN	Drawing Change Notice
DCS	Distributed Control System
DI	Digital Input
DIR	Direct
DO	Digital Output
DP	Differential Pressure
EPC	Engineering – Procurement – Construction
F	Fahrenheit
FC	Fail Closed
FDA	Food and Drug Administration
FI	Fail Indeterminate
FL	Fail Last, or Locked
FM	Factory Mutual
FO	Fail Open
FOB	Freight On Board
I&C	Instrumentation and Control
I/O	Input/Output
I/P	Current to Pneumatic (converter)

Abbreviation	Explanation
IA	Instrument Air
IEC	International Electrotechnical Commission
IEEE	Institute of Electrical and Electronics Engineers
IL	Instrument List
ISA	ISA - The International Society of Automation
ISA Dictionary	*The Automation, Systems, and Instrumentation Dictionary, Fourth Edition*
ISA-FG15	ISA Training Course FG15, Developing amd Applying Standard Instrumentation & Control Documentation, Version 2.2
ISA-S106	ISA Training Course S106, Version 2.0, Introduction to Measurement and Control — Technology, Industries & People
ISO	International Organization for Standardization
KO	Knock Out
mA	Milliamperes
MW	Manway or Access Port
NEC	National Electric Code
NEMA	National Electric Manufacturers Association
NFPA	National Fire Protection Association
OSHA	Occupational Safety and Health Administration
OWS	Oily Water Sewer
PAS	Plant Automation System
P&ID	Piping and Instrumentation Drawing
PE	Professional Engineer (licensed)
PES	Programmable Electronic System
PFD	Process Flow Diagram
PID	Proportional-Integral-Derivative
PIP	Process Industries Practices, founded by Construction Industries Institute (CII)
PLC	Programmable Logic Controller
PP	Personnel Protection
PPA	Prepay and Add (freight costs)
Psi	Pounds per square inch
Psig	Pounds per square inch gauge
PV	Process Variable
REV	Reverse
RFQ	Request For Quotation
ROI	Return On Investment
SAMA	Scientific Apparatus Makers Association
SCADA	Supervisory Control and Data Acquisition
SIS	Safety Instrumented Systems
SLDC	Single Loop Digital Controller
SP	Set Point
TAPPI	Technical Association of the Pulp and Paper Industry
T&C	Terms and Conditions
UL	Underwriters Laboratory
V ac	Volts - alternating current
V dc	Volts - direct current

Appendix C
Typical ISA-TR20.00.01 Specification Form

The following page was excerpted from the CD of ISA-TR20.00.01-2006, *Specification Forms for Process Measurement and Control Instruments, Part 1: General Considerations.* It is the first page of the device specification for a displacer-type level transmitter or local controller, Form 20L2121 Rev 0. The CD also provides additional pages that form a "pick list" (not included in this book). The "pick list" offers a choice for each of the lines of the device specification. For example, for line 13, "process connection, nominal size and rating", it is possible to select any one of eight sizes ranging from 1½ inches to 8 inches, or 50 mm to 100 mm for the process connection size.

1	RESPONSIBLE ORGANIZATION	DISPLACER-TYPE LEVEL TRANSMITTER OR LOCAL CONTROLLER Device Specification	6	SPECIFICATION IDENTIFICATIONS
2	ISA		7	Document no
3			8	Latest revision | Date
4			9	Issue status
5			10	

11	BODY OR CAGE			
12	Body/Cage type			
13	Process conn nominal size		Rating	
14	Process conn termn type		Style	
15	Flange facing finish			
16	Lower conn location		Upper	
17	Gage glass conn nom size		Rating	
18	Gage glass conn type		Style	
19	Vent/Drain conn nom size		Rating	
20	Vent/Drain termn type		Style	
21	Head orientation/type			
22	Extension/Heat insulator			
23	Body/Cage material			
24	Bolting material			
25	Gasket/O ring material			
26				
27				
28				
29	SENSING ELEMENT			
30	Sensor type			
31	Sp gr Lower Range-Limit		URL	
32	Displacer diameter		Length	
33	Extension length			
34	Displacer material			
35	Torque tube/Spring matl			
36	Extension/Cable material			
37	Trim material			
38				
39				
40	CONNECTION HEAD			
41	Type			
42	Enclosure type no/class			
43	Signal termination type			
44	Cert/Approval type			
45	Enclosure material			
46				
47				
48				
49	TRANSMITTER OR CONTROLLER w/wo SWITCHES			
50	Housing type			
51	Output signal type			
52	Enclosure type no/class			
53	Control mode			
54	Digital communication std			
55	Signal power source			
56	Contacts arrangement		Quantity	
57	Transient protection			
58	Integral indicator style			
59	Signal termination type			

60	TRANSMITTER OR CONTROLLER Continued			
61	Cert/Approval type			
62	Mounting location/type			
63	Span-Zero adjust lct			
64	Enclosure material			
65				
66				
67				
68	PERFORMANCE CHARACTERISTICS			
69	Max press at design temp		At	
70	Min working temperature		Max	
71	Accuracy rating			
72	Level Lower Range-Limit		URL	
73	Min level span		Max	
74	Min differential sp gr			
75	Min ambient working temp		Max	
76	Contacts ac rating		At max	
77	Contacts dc rating		At max	
78				
79				
80				
81	ACCESSORIES			
82	Connecting cables length			
83	Air set filter style			
84	Air set gauges			
85	Gage glass style			
86	Gage valve style			
87				
88				
89	SPECIAL REQUIREMENTS			
90	Custom tag			
91	Reference specification			
92	Special preparation			
93	Compliance standard			
94	Construction code			
95	Software configuration			
96				
97				
98	PHYSICAL DATA			
99	Estimated weight			
100	Overall height			
101	Removal clearance			
102	Upper to lower conn lg			
103	Lower to drain conn lg			
104	Signal conn nominal size		Style	
105	Mfr reference dwg			
106				
107				
108				

110	CALIBRATIONS AND TEST		INPUT OR SETPOINT			OUTPUT OR SCALE	
111	TAG NO/FUNCTIONAL IDENT	MEAS/SIGNAL/TEST	LRV	URV	ACTION	LRV	URV
112		Level-Analog output					
113		Level-Digital output					
114		Level-Scale					
115		Level setpoint 1-Output					
116		Level setpoint 2-Output					
117		Test pressure					

118	COMPONENT IDENTIFICATIONS		
119	COMPONENT TYPE	MANUFACTURER	MODEL NUMBER
120			
121			
122			
123			
124			
125			

Rev	Date	Revision Description	By	Appv1	Appv2	Appv3	REMARKS

Form: 20L2121 Rev 0

Appendix D
Drawing Sizes

The tables below compare architectural and engineering drawing sizes. Until 1995, sizes A through E, below, were used in the United States for both architectural and engineering drawings.

Engineering Drawing Sizes		
ANSI Size	**Dimensions (inches)**	**Dimensions (millimeters)**
A	8.5 × 11	216 × 279
B	11 × 17	279 × 432
C	17 × 22	432 × 559
D	22 × 34	559 × 864
E	34 × 44	864 × 1,118

Since 1995, the sizes ARCH 1 through 6 below have been used for Architectural Drawings.

Architectural Paper Sizes		
Size	**Dimensions (inches)**	**Dimensions (millimeters)**
ARCH 1	9 × 12	229 × 305
ARCH 2	12 × 18	305 × 457
ARCH 3	18 × 24	457 × 610
ARCH 4	24 × 36	610 × 914
ARCH 5	30 × 42	762 × 1,067
ARCH 6	36 × 48	914 × 1,219

Appendix E
Recommended References

The authors asked several I&CS practitioners what books they have near their desk for ready reference. We also have added our ideas to produce the following list.

Control Valves

- *Control Valves*, Guy Borden, Jr., Editor, Paul G. Friedmann, an ISA publication
  An overall view of control valves – their application, construction, design, and maintenance. The information assists engineers during control valve selection and operation.

- *Control Valve Primer, Fourth Edition,* an ISA publication, H. D. Baumann, an ISA publication
  This book gives practical advice on how to apply control valves. It presents insights on valve sizing, smart valve positioners, field-based architecture, network system technology, and control loop performance evaluation.

- *The Control Valve Handbook*, Fisher Controls, Second Edition, Third Printing
  There are several good manufacturers' handbooks. This one happens to be the one with which we are most familiar. It is a good reference that explains the function, features, terminology and individual components that comprise a control valve. The handbook explains how a control valve works and may provide insight into why your valve is behaving oddly. Even old timers will revisit the handbook from time to time.

Electrical

- **National Fire Protection Association (NFPA) Standard 70, National Electric Code (NEC) (latest edition)**
  Well, you can't, or should not, play the game without knowing the rules, and here are the rules, electrically anyway. There are good classes out there that will introduce neophytes to "the code", as well as offer refresher courses to the rest of us.

Flow

- *Flow of Fluids through Valves, Fittings and Pipe*, Crane Technical Paper 410, 1985, The Crane Company

  This is that little orange spiral bound book seen on control systems engineers' desks, as well as on the desks of process and mechanical engineers. Crane Technical Paper 410 contains an invaluable memory jogging multi-page listing of fluid flow equations, an explanation of how to calculate fluid flow losses through valves and fittings, fluid property tables, and even tables showing conversion factors. It is probably the most used reference book on many desks.

- *Flow Measurement Engineering Handbook*, Third Edition, R.W. Miller, McGraw-Hill

  This is a fine handbook providing clear, concise, and, if you want it, detailed and technical information about the science and application of flow measurement devices. It is a good source for orifice plate bore calculations. We recommend that everyone run through the manual calculation occasionally so you can better appreciate the software. It also is good to remember the factors that influence the reading you see after the system is installed.

- *Flow Measurement*, Second Edition, D. W. Spitzer, Editor, an ISA publication

  An excellent handbook covering the whole array of flowmeters. Emphasis is placed on the accuracy of measurement and how to ensure it. The book includes topics on forecasting flow conditions, flowmeter selection, installation, calibration, and maintenance.

Standards

- ANSI/ISA-5.1-2009, *Instrument Symbols and Identification*

  This is the oracle for anyone documenting a control system. From P&IDs to sketches, this is the standard for the symbolism used. This was the source document for thousands of P&ID legend sheets – and hundreds of thousands of spirited discussions known colloquially as arguments.

- ISA-5.4-1991, *Instrument Loop Diagrams*

  So what makes a Loop Diagram? Well, this standard not only establishes the content of the ubiquitous loop, but it also presents increasing levels of complexity that can be seen in typical Loop Diagrams. ISA-5.4 includes sample loop diagrams for pneumatic, electronic and shared display systems. The sample loops define the minimum requirements for Loop Diagrams as well as additional optional items that can be included.

- ISA-20-1981, *Specification Forms for Process Measurement and Control Instruments, Primary Elements, and Control Valves*

  Not only are there blank specification data sheets in this standard, but next to each data sheet template, a line-by-line listing can be found telling the user what information each block or datum was supposed to provide.

Temperature

- ***The Temperature Handbook*, 27th Edition (or latest), Omega Technologies**

  Almost everyone has at least one Omega handbook at their desk. The one referred to most often might be *The Temperature Handbook*, as it includes so much information about thermocouples, RTDs and other temperature elements, transmitters and controllers – "hey, what color code is used for a type K thermocouple, anyway?". Contains easy-to-understand technical information on many topics.

Vendors

- **Vendor catalogs, CDs, or more commonly, Web access to catalog information.**

  If you are working with instrumentation, Internet access to manufacturer's current specification information and model numbers is essentially a requirement. Some manufacturers have well organized, intuitive and accessible Web sites; some have simply awful Web sites. Thankfully, the better instrument and control valve suppliers have user-friendly access to ALL their technical data, which we feel is a critical criterion for instrumentation and controls professionals. If your supplier doesn't have accessible, current and complete information available on the Web, including installation and termination information, ask them to fix the site or think about finding another supplier. Hopefully, all our suppliers will eventually respond to suggestions, so it is up to us to share the pain when the sites don't work like we need them to work. The worst offenders offer minimal, glossy information with no technical "meat", telling us to contact our local representative for details. In our experience, it is better to simply use a different supplier. Many instrument users link to vendor Web sites or have loaded the manufacturers' installation, operation and maintenance manuals onto their Intranet, so there are no libraries of out-of-date information to lead us astray.

- ***ISA Directory of Automation***

  The *ISA Directory of Automation* is an online directory, produced in collaboration with MultiView, Texas, USA, offering extensive information about manufacturers and service companies at your disposal whenever you need it.

Index

A

ANSI/ISA-5.06.01-2007, Fig. 10.5 195
ANSI/ISA-84.00.01 103-106, 153
ANSI/ISA-88.00.01 17, 153-154
API 154
Area Classification, Electrical, Fig. 4-8 78-81
ASME 154
ASME Y14.1, Fig. 9-1, Fig. 9-2 137-141

B

Batch Processes 1, 17
Bid 85-90
- Evaluation, Fig. 5-1
- Package

Binary Logic 123-133
- Implementation, Fig. 6-1 - Fig. 6-9
- Symbols, Fig. 6-6, Fig. 6-7
- Systems, Fig. 6-1 – Fig. 6-9

C

Company Standards 150-151
Comparing, Fig. 2-1, Fig. 2-2 21-22
Consensus Standards 151-156
Continuous Processes Intro
Control Loop, Fig. 2-4 23
Control Valve
- Failure Position, Fig. 2-5 24-25
- Symbols, Fig. 2-5, Fig. 2-16 39

Controlled Documents 20-21
Controller
- Direct Acting, Fig. 7-5 113-115
- Reverse Acting, Fig. 7-5 113-115

Correcting, Fig. 2-3 22-23

D

Database, Fig. 3-2, Fig. 3-3
Deliverables 4, 10
Direct-Acting Controller, Fig. 7-5 113-115
Discrete Control 9
Discrete Manufacturing Processes 1
Drawing
- Bills of Material 123-125
- Change Notice Notes, Fig. 9-5
- Numbers 141-142
- Title Blocks, Fig. 9-3, Fig. 9-4 140-144
- Revisions, Fig. 9-4, Fig. 9-5 144-145
- Sizes, European, Fig. 9-2 139
- Sizes, U.S., Fig. 9-1 137-138

E

Electrical Schematic Drawings, Fig. 6-4
Electronic Loop, Fig. 7-2 108
Electronic Transmission, Fig. 2-15 38-39
Engineering Design Team 14
Equipment
- Identification
- Symbols, Fig. 1-3 15-16

European Drawing Sizes, Fig. 9-2 139

F

FDA 147-148
Flammable Materials
Flow
- Devices, Fig. 2-17, Fig. 2-18 54
- Loop, Fig. 2-22 63
- Transmitters, Fig. 2-1 31

Functional 96
Diagrams, Fig. 6-2, Fig. 6-3
Specifications 96

G

Government Standards 147-150

H

Hazardous (Classified) Locations, Fig. 4-7, Fig. 4-8 78-81

I

IEC 156
Industry Standards 150-151
Installation Details, Fig. 8-1, Fig. 8-2, Fig. 8-3 123-136
Instrument
Database 55-66
Definition, Fig. 2-1 31
Drawing Schedule, Fig. Intro 4 7
Index/List, Fig. 3-1 55-66
Installation Details, Figs 8-1, 8-2, 8-3 123-136
Loop, Figs. 2-4, 2-22, 2-23, 2-24 34, 63, 64, 65
Tag Number, Fig. Intro-3 8, 9, 19-54
Instrumentation and Controls
Design Group 2
Documents
ISA-5.1, Figs. 2-5, 2-7, 2-10, 2-12, 2-13, 2-16 thru 2-18, 10-5 19-54
ISA-5.2, Fig. 10-5 152-154
ISA-5.3, Fig. 10-5 152-153
ISA-5.4, Figs. 7-3, 7-4, 7-5, 7-7, 7-8, Fig. 10-5 107-122, 152-153
ISA-5.5, Fig. 1-3, Fig. 10-5 15, 16, 152-153
ISA-d5.07.01, Fig. 10.5
ISA-20, Fig. 4-2 thru Fig. 4-6 67-82
ISA-TR20.00.01, Appendix C 67-82
Intrinsic Safety
ISA
International Society of Automation
Standards 147-156
ISO 156

L

Ladder Diagrams 95, 97-98
Level
Instrument Specification Form, Fig. 4-4, Fig. 4-5, Fig. 4-6 73-78
Loop, Fig. 2-24 50-51
Line Symbols
Location Plans, Fig. 8-4, Fig. 8-5, Fig. 8-6 7, 123-136
Logic Diagrams, Fig. 6-9 6, 95-122
Loop Diagrams, Fig. 7-3 thru Fig. 7-8 6, 107-122

M

Mandatory Standards
Food & Drug Administration 147-148
National Electric Code 148
OSHA, Fig. 10-2, 10-3, 10-4 149-150
Mechanical Completion 121

N

NEC 148
NEMA
NFPA, Fig. 10-1

O

On-Off Control 9, 95-106
OSHA, Fig. 10-2, Fig. 10-3, Fig. 10-4 149-150

P

P&ID
Piping and Instrumentation Drawing, Fig. Intro-4, Fig. 2-21 6-7, 19-53
Symbols, Figs. 2-5, 2-6, 2-8, 2-9, 2-12, 2-13, 2-16 thru Fig. 2-20 19-56
Payment Terms 92
PIP 155
Piping Design Group 8, 125-126, 132-133
Pneumatic Loop, Fig. 7-1, Fig. 7-6 108, 116
Pneumatic Transmission, Fig. 2-14 37
Pressure
Gauge Instructions, Fig. 4-3 70-73
Gauge Specification Form, Fig. 4-2 70-73

Loop, Fig. 2-4 34
Process
Control (Definition), Fig. 2-1 31
Control Description, Fig. 2-1
Description, Fig. 1-2 13-16
Design Documentation, Fig. INTRO-4 1-8
Diagram
Drawing Schedule, Fig. INTRO-4
Process Flow Diagram, Fig. 1-1 5, 7, 11-18
Process Industry
Processes
Batch, Fig. 1-4 1, 17
Continuous 1
Discrete Manufacturing 1
Purchasing, Fig. 5-1 83-94

R

Recipes, Fig. 1-4 17
Recommended Practices (ISA) 152
Reference Table
Regulations, Fig. 10-1 thru Fig. 10-4 147-157
Reverse Acting Controller, Fig. 7-5 113-115
Revisions, Fig. 9-4, Fig. 9-5 144-145

S

SAMA 155-156
Sensing, Fig. 2-1, Fig. 2-2 31-32
Shipping 91-92
SIS
Safety Instrumented Systems, Fig. 6-11 104-106
Separated, Fig. 6-11 104-105
Specification Forms, Fig. 4-1 thru Fig. 4-6 6, 67-82,
Standards
Consensus 151-157
Company 150
Government 150
Industry 150
ISA 151-154
Mandatory, Fig. 10-1 thru Fig. 10-4 147-150

T

Tagging 89
TAPPI 156
Title Blocks, Fig. 9-3, Fig. 9-4 140-144
Transmission
Electronic, Fig. 2-15 52
Pneumatic, Fig. 2-14 51

U

U. S. Drawing Sizes, Fig. 9-1 137-138, Appendix D